ENVIRONMENTAL HEALTH -
PHYSICAL, CHEMICAL AND BIOLOGICAL FACTORS

RURAL POVERTY AND DEGRADATION OF NATURAL RESOURCES IN GHANA

ENVIRONMENTAL HEALTH - PHYSICAL, CHEMICAL AND BIOLOGICAL FACTORS

ENVIRONMENTAL HEALTH -
PHYSICAL, CHEMICAL AND BIOLOGICAL FACTORS

RURAL POVERTY AND DEGRADATION OF NATURAL RESOURCES IN GHANA

ALEX OBENG SOMUAH
AND
EDWARD DEBRAH WIAFE

NOVINKA
New York

Library of Congress Cataloging-in-Publication Data

Names: Somuah, Alex Obeng, author. | Wiafe, Edward Debrah, author.
Title: Rural poverty and degradation of natural resources in Ghana / Alex Obeng Somuah and Edward Debrah Wiafe.
Description: Hauppauge, New York : Nova Science Publishers, Inc., [2016] | Series: Environmental health - physical, chemical and biological factors | Includes bibliographical references and index.
Identifiers: LCCN 2016021562 (print) | LCCN 2016033784 (ebook) | ISBN 9781634854085 (softcover) | ISBN 9781634854245 ()
Subjects: LCSH: Poverty--Environmental aspects--Ghana. | Rural poor--Ghana. | Land degradation--Ghana. | Ghana--Economic conditions. | Ghana--Environmental conditions.
Classification: LCC HC1060.Z9 P627755 2016 (print) | LCC HC1060.Z9 (ebook) | DDC 333.71/3709667--dc23
LC record available at https://lccn.loc.gov/2016021562

Published by Nova Science Publishers, Inc. † New York

CONTENTS

PREFACE

Standard of living of human beings can influence the extent to which natural resources are being utilized and managed. The proximity of people to the natural resources, increase accessibility to and abuse of their use especially where there are no control. Degradation of natural resources in Ghana has been strongly influenced by the standard of living. Poverty as a living condition has been regarded as a major factor causing degradation of natural resources. The situation even deepens when the poverty occurs in the rural areas of developing countries and on marginal lands where extreme weather conditions aggravate.

In this book, the quest to exploit natural resources as a means of surviving in poverty of soil has been demonstrated to have degraded some natural resources. It further provides a significant contribution towards systematic evaluation of poverty and regulates measures to exploit natural resources sustainably for survival. A case study has been presented in the book to demonstrate the relationship between natural resources degradation and poverty in rural area. It has been assumed that the society wishes to conserve natural resources for whatever reasons, economical, ecological, aesthetic, and cultural or any combination of these. However, this assumption might not be valid always as there are individuals in the society who would rather overexploit the resources for their selfish gains. The serious resource manager cannot ignore the issues raised in the book especially the relationship between poverty and natural resource degradation in sub-Saharan Africa. Whether the society manages the natural resources sustainably or waiting for the government to manage it for them is left for the public scrutiny.

ACKNOWLEDGEMENTS

It will be necessary to give credit to all those whose contributions have complemented our efforts. We first, wish to express our sincere thanks to Professors S.B. Kendie, J.V. Mensah, Mansah Prah and Kumi Kyereme all of University of Cape Coast, Ghana; for their encouragement and guidance. Again, we cannot forget the immeasurable sacrifices and services of the following organizations: Kwahu South District Assembly, Holy Family Hospital, Nkawkaw; Chiefs and people of Kwahu Traditional Area for their contribution to the data collection. We are also indebted to Professor Amuaquando, Mr. Imoro Rasak, Mr. Sika Bright and Ms. Esther Nana Ama Yeboah for their support during the data collection, analysis and editorial issues. We would be unforgiveable to forget the contributions of Rev. Professor Addow Obeng, the President of the Presbyterian University College, Ghana.

LIST OF TABLES

LIST OF FIGURES

LIST OF PLATES

Chapter 1

INTRODUCTION

The growth of societies depends on land and how societies make use of it to eke out living. Therefore, societies in their quest for survival have passed through a number of stages to the present stage. Societies started from hunting and gathering characterised by collective living. There were low level of division of labour and specialisation was based on age and sex. This system of survival depended solely on the natural environment.

The discovery of agriculture ushered in sedentary life and enabled accumulation of surplus food. The dependence on land and the environment which was further propelled by the discovery and use of metal tools increased due to the quest to increase productivity. The stage of human society that saw greater and more mechanized dependence on land was the industrial revolution dominated by detailed division of labour.

OVERVIEW OF ENVIRONMENTAL DEGRADATION AND SOME ASSOCIATED THEORIES

Among the varied challenges faced by humankind today, the biggest are food shortages and environmental degradation, both resulting from population explosion and poor resource management [1]. Deforestation in the tropics is often driven by agricultural expansion and exploitation of forest resources [2]. Land clearing for agricultural expansion takes the form of burning or one-time logging. Burning, decomposition of waste forest matter, soil degradation in cleared lands, and other land use activities annually emit approximately 1.6 GtC into the atmosphere, largely as CO_2 [2].

Logging accounts for 20 percent of global deforestation, roads and infrastructure construction account for 15 percent, and cattle ranching accounts for 10 percent. These indicate the extent of land degradation globally. In addition, the world tropical forest is believed to contain 40-50 percent of all the animals and plant species in the world and includes many endemic species [3].

Despite the survival value of the forest to humanity, it is becoming obvious that human beings have engaged in unsustainable use of the forest leading to the loss of huge tracts of the tropical forest. The Earth's tropical forests were once 40 percent rainforest and 60 percent dry forest. Today, agriculture and anthropogenic fires have essentially obliterated dry forests in the tropics. Tropical forests have lost half of their original expanse in the past 50 years, the fastest vegetation change of this magnitude in human history. A century ago, the tropical forest covered about 12 percent of the world's surface but now it covers only 5 percent [4].

A study using globally comprehensive and consistent satellite data estimated that the extent of the world's remaining closed natural forest (where crown cover is more than 40 percent) in 1995 was 2,870 million ha, about 21.4 percent of the land area of the world [5]. It has been estimated that 25 percent of the species which existed in the mid-1980s may be extinct by 2015. If tropical forest conversion continues, the world may lose between 5 and 15 percent of its total plant and animal species by 1990 and 2020 respectively [6]. The rates of deforestation in Africa were about 18,947.8 hectares and 11,903.1 hectares per year in the 1980s and 1990s respectively [7].

With respect to the rates of tropical deforestation mentioned above, United Nations Educational, Scientific and Cultural Organisation [8] argued that in tropical Africa about 8,500 plant species and 403 species of orchids have been noted down e.g., the pygmy hippopotamus (*Hexaprotodon liberiensis*), the giant forest hog (*Hylochoerus meinertzhageni*), the water chevrotain (*Hyemoschus aquaticus)* and a number of insectivores, rodents and bats, tree frogs, bird species inhabit the area. These species, along with a diversity of fruits and insects, make a special habitat. Top canopy monkey species, the Miss Waldron's red colobus (*Piliocolobus badius waldronae*), and others, already have disappeared from much of tropical Africa's forest all due to deforestation. These revelations have significant economic and social implications. More than three-quarters of the world's population depend directly on plants for food and medicines extracted from the forests [9].

Forests can serve as subsistence safety nets for the rural poor, essentially mitigating poverty for its users [10]. Forests can also function "as a source of permanent increases in income, assets, services, civil and political rights, voice, and the rule of law" (ibid.), particularly in well-functioning community-managed forests. This includes contributions to local education, clean water, infrastructure, provisions for health and nutritional programs, etc.

Community forestry projects tended initially to focus on justice (improving access to forest for traditional users) and on the hypothesis – well proven in the event – that it made sense to protect forest with, and not through conflict with, its users. Some people assumed that forests would usually be better protected by their users than by the State, and that hypothesis too is largely proven.

Top soils in temperate forest may be two metres deep than in the tropical forest which may rarely exceed five centimetres [11]. Thus, if the tropical forest is damaged, it cannot repair itself, since the heavy rains wash the nutrients away and the hot sun bakes the thin layer of top soil so much that only coarse grass can grow.

The world's population is increasing at an annual rate of more than 90 million people. This requires additional land resources for food, shelter and energy, which has direct impact on natural resources. If current trends in population growth and depletion of forest resources remain unchanged, the number of people that experience fuel wood shortage is expected to increase from 1500 million in 1980 to 2400 million in the year 2000 [12, 13, 14]. The literature demonstrates clear link between rapid population growth and declining natural resource base leading to increased poverty. Its problems have been studied by various disciplines including economics, ecology and geography.

The Malthusian perspective attributes decline in resource base and poverty to the ever increasing population. More than 90 percent of African agricultural production is estimated to come from small-scale producers. At least 90 percent of the world's poor have been estimated to be dependent on forests for at least some of their income [15].

However, the Boserup school of thought maintains that increasing population leads to development through increased agricultural innovation and intensive land uses [16, 17, 18] argued that slowly increasing population densities have desirable effects on technical change, land and labour productivity and rural per capita income increases through changes in relative factor prices.

Another argument of the Boserupian school of thought, according to [19], is that the faster the improvement in factor productivity, the smaller will be the proportion of land and population that will be engaged in agriculture and the greater the possibility that increasing land areas can be left fallow or reforested.

There are many problems with the population – environment link. For instance, the conditions (technological, economic and rate of change of population densities) under which agricultural intensification may take place with increasing population densities are not uniform worldwide. For many areas of the developing world, the reality is one of rapid population growth leading to the parceling of agricultural lands, cultivation of marginal lands, migration and increased poverty. Owing to population numbers and changes in tastes and preferences, people are going to make more demands on the environment. The latter, which is associated with affluence, puts further pressure on the environment because of the need to increase consumption.

For instance, with increasing affluence and change in tastes, the demand for animal products, as opposed to cereals, increases. This may have to be satisfied either through expansion in agriculture or through imports. In countries such as Ghana, the demand for foreign exchange to meet imports puts further pressure on environment-based resources such as minerals, timber and agricultural produce which have for decades, been the major foreign exchange earner for the country. Therefore, the opportunities offered by population [density] change to improve land productivity as the Boserupians hold cannot be actualized in many areas. One reason is that in the developing world, rapid population growth is often the result, not the cause, of social and economic inequality [20].

Some theorists view over population as the central cause of environmental problems, while others emphasize the role of harmful technological innovation.

From a conflict perspective, however, neither of these approaches is satisfactory [21] nor draws on the conflict perspective and blames the dominant social paradigm of advanced capitalist societies for threatening the environment.

Greater share of the human and natural resources of the developing countries is being appropriated to the industrialized nations. This process only intensifies the destruction of natural resources in poorer regions of the world. From a conflict perspective, less affluent nations are being forced to exploit their mineral deposits, forests and fisheries in order to meet their debt obligations. The poor turn to the only means of survival available to them: they plough mountain slopes, burn plots in tropical forests and over graze grasslands [22]. This interplay between economic troubles and environmental destruction has been quite evident in Brazil, where each year, thousands of square kilometres of the Amazon rainforest are cleared for crops and livestock through burning. The elimination of the rainforest affects worldwide weather patterns and heightens global warming.

Environmental destruction in Southern Honduras is intricately connected to problems of poverty, unemployment and land tenure as has been suggested by [23]. Stonich found that the policies of multinational corporations and international aid organisations encourage an export-oriented agriculture that relies on intensive land use. In her view, such policies are likely to lead to destruction of the remaining rain forests in Central America while increasing inequality, malnutrition and conflict within and between nations. These socio-economic patterns with harmful environmental consequences are evident not only in Latin America but also in many regions of Africa and Asia. Nevertheless, as we are well-aware of the environmental implications of land use policies in the Third World, the conflict theorists contend that such a focus on the developing countries can contain elements of ethnocentrism.

Conflict theorists point out that western industrialized nations account for only 25 percent of the world's population but are responsible for 85 percent of worldwide consumption. Looking at the United States alone, a mere six percent of the world's people consume more than half the world's non-renewable resources and more than one-third of all the raw materials produced. With such evidence in mind, conflict theorists charge that the most serious threat to the environment comes from affluent mega consumers and mega polluters.

Schnaiberg (1980) later criticized the focus on affluent consumers as the cause of environmental troubles. In his view, a capitalist system has a "treadmill of production" because of its inherent need to build over-expanding profits. This treadmill necessitates creating an increasing demand for products, obtaining natural resources at minimal costs, and manufacturing products as quickly and cheaply as possible, no matter the long-term environmental consequences of this treadmill. In his view, it is those who own and control the means of production that are ultimately responsible for the effect of this treadmill of production resulting in ever-expanding consumption.

Large parts of sub-Saharan Africa have distinctive wet and dry seasons, which favour regular vegetation fires. The wet season stimulates growth, while the dry season provides ideal conditions for burning. Droughts increase the possibility of wild fire, due to extreme poor weather conditions. In addition, every year the El Nino weather pattern provides climatic conditions that favour extended wildfire episodes [24].

According to [24], Africa is the most fire-prone continent. An estimated 2.3 million kilometres or 7.7 percent of the continent was burnt in 2000. Africa also had the highest number of fires (54 percent). Burning in 2000 was most extensive in East Africa (873840 km^2 or 15 percent of the area) Central Africa (539,225 km, 13.5 percent and Southern Africa (677,123km^2, 11.5 percent).

Ghana's environment is suffering from the effects of climatic change resulting from the destruction of its forests. Most of the forests have been degraded into Savannah, and some forest vegetation are so impoverished making it difficult to regenerate naturally [25]. Today, forest estates continue to dwindle at an estimated annual deforestation rate of 2% (65,000 ha), valued at annual degradation cost of 10% of Gross Domestic Product (GDP) [26].

Indicators of Poverty

According to the [27], it has been estimated that in 2007, more than 80 percent of the world's population lived in countries where income differentials were widening. The poorest 40 percent of the world's population accounts for 5 percent of global income and a mere 12 percent use 85 percent of its water and this 12 percent do not live in the developing world. Some 1.1 billion people in the developing countries have inadequate access to water and 2.6 billion lack basic sanitation. Nearly 1 billion people entered the 21st century unable to read a book or sign their names [27]. In Sub-Saharan Africa extreme poverty went up from 41 percent in 1981 to 46 percent in 2001, which combined with growing population increased the number of people living in extreme poverty from 231 million to 318 million. From the same Report, nearly a billion people are illiterates with about 800 million people lacking access to health services. Over a billion people lack access to safe water while some 840 million go hungry or face food insecurity.

They [28] also found out that malnutrition indicators are worse for the poor and predominant for the developing world. In year 2000, approximately 2.9 billion people, more than half the people in developing countries lived on less than $700 a year. Out of these, 1.2 billion earned less than $1 a day, with 2.8 billion earning $2 a day. Furthermore, in a minute, more than one woman dies during childbirth. Malaria, affects an estimated 300 million people and kills at least one million people each year, three-quarters are children under five years. Furthermore, 325 million children were out of school in 2000 at the primary and secondary levels, and of this 183 million were girls. It is estimated that 100 million school-age children will not be in school by 2015.

In 2000, 854 million adults were illiterate, of these 64 percent (543 million) were women [29].

Poverty is very predominant in the less developed countries. South Asia has the largest number of poor people (490 million) but there is a steady concentration of poverty in Africa [29]. Africa is the only continent where poverty is rising [29]. From various reviews and reports on the economic and poverty situation in Sub-Saharan Africa (SSA), it was concluded that the region faces difficult task to meet some, if not all, of these goals by 2015 [30]. About 300 million people live in poverty in Africa and the World Bank estimates that African economies would need to grow by five percent a year just to keep this number from rising. If the number of people living in extreme poverty is to be halved by 2015, as called for under the Millennium Development Goals [MDGs], not only would economies need substantial growth of at least seven percent, but incomes would also have to be distributed more equitably.

An analysis of survey data for Africa assembled by [31] reveals that in 94 per cent of 115 'poverty line' studies, rural poverty was recorded at higher levels than urban poverty. For 65 per cent of cases the rural- urban poverty ratio was in excess of 1.5:1 (i.e., rural poverty levels were more than 50 per cent greater than urban levels). In absolute terms these percentages translate to 633 million in Asia, 204 million in Sub-Saharan Africa, 27 million in the Near East and North Africa and 76 million in Latin America and the Caribbean.

Insufficient information is available concerning the extent to which trees surviving logging are sufficiently damaged that they die some time later, for example through fungal infection of wounds, but some kind of damage to 40% of trees in the remnant stand is common [32]. Many countries in Africa face exceptional food shortages, and millions of people still face the threat of famine and starvation. Even without famine and starvation, malnutrition is widespread. According to the [33], just under half of the region's population suffers from some level of food deprivation, with serious consequences for health and productivity.

According to the [34], Ghana's population is about 18.5million, indicating a tripling of the population between 1957 and 2000 or doubling of the 1970 population in just 26 years. This gives an estimated population growth rate of 2.8 percent. Between 2000 and 2010, Ghana's population increased from 18.5 million to 24.3 million. This high rate of growth is a source of worry for policy makers and planners. The fertility rate of 5.5, life expectancy of 57 and infant mortality rate of 66 per 1000 and maternal mortality of 216 per 100,000 live births (Ghana Living Standards Survey [GLSS] 4) are issues of great concern to a developing country like Ghana.

People perceive poverty as a threat to the very existence of humankind. Unequal distribution of global wealth has exacerbated the problem of poverty in the developing world. This high population has to be supported by the environment. It is therefore not surprising that Ghana, with a forest area of 8.2 million hectares in 1957 now has only 1.6 million hectares of forestland. Ghana has mismanaged and misused the environment to the extent that the country is now battling with desertification, poor and unreliable rainfall patterns.

In terms of international ranking of poverty, Ghana ranks barely above the least developed nations of the world. About one-third of all Ghanaians are classified as poor [35]. Since the United Nations Development Programme launched the Human Development Report in 1990, Ghana has continuously ranked among those countries with low human development. Ghana ranked 138th out of 177 in 2005 and 136th out of 174 countries in 2006 and 135th out of 174 countries in 2007. An examination of social conditions in Ghana shows generally poor welfare indicators. Of the three anthropometric indicators of malnutrition (stunting, wasting and underweight), stunting is the most prevalent among the children aged 0-4 years. Nearly one third (32.4 per cent) of the children under the age of five years are stunted (short for their age) compared to 15.5 per cent for wasted (underweight for age for height) and 25.8 per cent for underweight (underweight for their height for age). On access to health facilities in rural communities in Ghana only 27 percent of the rural poor live within 30 minutes to a health facility [36].

According to [36], about 31 percent of all adults have never been to school, less than one-fifth (17.1%) attended school but did not obtain any qualifications; 39 percent have Middle School Leaving Certificate [MSLC]/ Basic Education Certificate Examination [B.E.C.E] and National Vocational Training Institute [NVTI] certificate as their highest qualification, while a small percentage of 13.6 possess secondary or higher qualification. It is indicated that only 50 percent of adults in Ghana are literate in English or a local language with a little over 6 out of every 10 men, but fewer than 4 out of every 10 women, being literate. Almost 66 percent of adults in urban areas are literate but in the rural areas, only 41 percent are literate. Also, only 42 percent of households have access to pipe-borne water with as much as 57 percent of the households in Ghana using wood as their main source of fuel for cooking.

About 20 percent of households do not have access to toilet facilities with 27 percent in rural forest zone and 70 percent in rural savannah respectively having to ease themselves in the bush or the beach [free range]. Only 18.8 percent of rural dwellers enjoy pipe borne water as against 80.3 percent for urban areas. The illiteracy situation in Ghana stands at 45.9 percent with rural illiteracy at 59 percent as against 34 percent for the urban areas. More than three-quarters of urban households have electricity for lighting, compared to only 17 percent of rural households. Most urban households use charcoal for cooking, whereas most households in the rural areas use firewood. Only 14 percent of urban households and 2 percent of rural households have access to a flush toilet [34].

The [34] further shows that apart from Greater Accra (87.7 percent) and Ashanti (51.3 percent), the rest of the country remains predominantly rural, in spite of the substantial increase in the level of urbanization since 1984 (43.8 percent in 2000 as compared to 32 percent in 1984). None of the remaining regions has a level of urbanization that is above the national average (43.8%). This indicates that a significant part of the country can be described as rural. Poverty in Ghana is both a rural and urban phenomenon, however it is more pronounced in the rural areas.

Chapter 2

CONCEPTUAL DEFINITION OF POVERTY AND LAND DEGRADATION

DEFINITION OF POVERTY

Sociologists and other academics have found it extremely difficult to give it a universally acceptable definition. The perception of poverty has evolved historically and varies tremendously from culture to culture. Criteria for distinguishing poor from non-poor tend to reflect specific national priorities and normative concepts of welfare and rights. In general, as countries become wealthier, their perception of the acceptable minimum level of consumption, (i.e., the poverty line) changes [37].

Poverty is increasingly viewed as a multidimensional concept. It has social and psychological effects that prevent people from realizing their potential. Measurement of poverty can include material deprivation, isolation, alienation, dependence, and lack of participation or freedom of choice of assets, vulnerability and insecurity. Measurement is based on material deprivation generally linked to the inability of incomes to meet basic nutritional demands. Poverty is operationally defined as the inability to attain a minimal standard of living [38]. Poverty is mostly defined in two ways: absolute poverty and relative poverty.

The earlier attempts to define poverty were based on the view that a family required certain basic commodities if they were to survive. If they were unable to acquire these things, then they were in absolute poverty [39] describes absolute poverty as a situation in which people barely exist where the next meal may literally be a matter of life or death as the cumulative effects of malnutrition and starvation. Existence becomes a matter of struggle because people who find themselves in this kind of poverty do not have any hope of getting the next meal and therefore below or beneath this level of existence is only death.

[40] Says absolute poverty represents the problem of poverty in its rawest form. It means lack of food, cash and any assets. It is a 'condition' of life so degraded by disease, illiteracy, malnutrition and squalor as to deny its victims basic human necessities. It thus means a condition in which it is not possible to obtain the basic needs of life or where deprivation is so severe that the basic needs of life can scarcely be met at the minimum level of survival. In other words, absolute poverty is the state of existence in which the overall needs of the individual are not satisfied due to lack of enough purchasing power or means for self-provisioning.

With relative poverty, people essentially tend to compare themselves with others in their reference group to which they think they may fairly be compared because of broadly similar characteristics [41] says individuals, families and groups in the population can be said to be in poverty when they lack the resources to obtain the types of diet, participate in the activities and have the living conditions and amenities which are customary, or at least widely encouraged or approved in the societies to which they belong. Their resources are so seriously below those commanded by the average individual or family that they are in effect excluded from ordinary living patterns, customs and activities [42] notes that the poor are those people who are deprived of the things other people take for granted. The poor can be defined as those whose income or resources fall so far short of the average within a society that they do not have an acceptable standard of living.

According to [43], the poor are those who are unable to obtain an adequate income, find a stable job, own a property, or maintain healthy living conditions. They also lack an adequate level of education and cannot satisfy basic health needs. The poor have no access to the basic necessities of life such as food, clothing and decent shelter; they are unable to meet social and economic obligations; they lack skills and gainful employment; have few, if any economic assets; and sometimes lack self-esteem [44].

The [45] committee on the study of poverty defines poverty as a dearth of key resources in terms of development objectives. Poverty is therefore, the lack of basic human necessities, a condition arising largely from total absence, scarcity or under-development of requisite resources.

From a human development perspective, poverty means the denial of opportunities and choices for a tolerable life [46, 47] asserts that it is the deprivations of the lives people live that poverty manifest itself. Poverty can mean more than a lack of what is necessary for material well-being. It can also mean the denial of opportunities and choices most basic to human development that is to lead a long, healthy, creative life and to enjoy a decent standard of living, freedom, dignity, self-esteem and respect from others.

[48] Identifies five types of rural poverty. Material deprivation and alienation cause interstitial poverty, or pockets of poverty surrounded by power, affluence and ownership of assets. Material deprivation can combine with isolation and alienation to lead to peripheral poverty, which is found in the marginal areas. Material deprivation arising from population pressure and limits on resources will breed alienation and overcrowding poverty. Vulnerability to natural calamities, (e.g., drought) labour displacement, and insecurity, produces traumatic or sporadic poverty, which can be transitory but often ends up being endemic. Isolation, alienation, technological deprivation, dependence and lack of assets are also signs of endemic poverty.

This classification is important for linking the types of poverty processes to the types of poverty produced and the segments of the population affected [48] indicates that the proportion of the populace living in family units with expenditure or income per individual below the poverty line has been on the decrease in the world's regions since 1990.

In Ghana, the Statistical Service in Ghana Living Standard Survey, 2005/2006 (GLSS 5) anchors the poverty line on the nutrition needs of the Ghanaian population. It derives two lines: a lower line of ¢2,884,700 cedis (GH¢288.47) per adult equivalent per year, and an upper line of ¢3,708,900 cedis (GH¢370.89) per adult equivalent per year. Therefore, 'the poor' for the study are defined as:

- Those living below a per capita annual income of ¢3,708,900 (GH¢370.89) and cannot meet their needs.
- Those who do not have access to household property such as, tape recorder, television, fridge furniture-bed/the simplest form of living room furniture, etc.
- Those who do not have access to productive assets such as land, spraying machine, commercial vehicle and house for renting.
- Those who do not have access to public services such as potable water, educational and health facilities, toilet facilities, sanitation and electricity.

According to GLASOD (Global Assessment of Soil Degradation) [49] land degrading occurs when its function is lost or its productivity is reduced in terms of quantity, quality, goods and services. A decline in productivity would be the most obvious sign of land degradation, but it would not be sufficient to establish degradation: a decline could have come about for purely social reasons.

DEFINITION OF LAND DEGRADATION

[50] Asserts that land degradation is a composite term. It has no single readily-identifiable feature, but instead describes how one or more of the land resources (soil, water, vegetation, rocks, air, climate, relief) has changed for the worse. Land degradation is used in the broad sense to mean impoverishment of the land by human activities and by natural causes (such as climate). Land is considered "degraded" when its productivity is diminished.

According to [51], a wide range of terms is used in the literature to define land degradation, often with distinct disciplinary-oriented meaning, and leading to misinterpretation among disciplines. Some common terms used are soil degradation, land degradation, and desertification.

According to [52], land degradation is the loss of utility or potential utility or the reduction, loss or change of features, organisms which cannot be replaced [53] see land degradation as a process which lowers the current and/or potential capability of soil to produce (quantitatively and/or qualitatively) goods and services. It distinguishes six main groups of degradation processes:

(1) water erosion (sheet, rill, gully, mass movement),
(2) wind erosion,
(3) excess of salts,
(4) chemical degradation (toxicity, acidification),
(5) physical degradation (loss of structure, sealing, crusting), and
(6) biological degradation (loss of biodiversity, general impoverishment, decreases of cover).

According to [54], land degradation implies reduction of resources potential by one or a combination of processes acting on the land. These processes include water erosion, wind erosion and sedimentation by those agents, long-term reduction in the amount of diversity of natural vegetation, where relevant, and salinization and sodication. Land degradation is the temporary or permanent lowering of the productive capacity of land [54]. It thus covers the various forms of soil degradation, adverse human impacts of water resources, deforestation and lowering of the productive capacity of range lands.

Soil fertility decline is used in the short term to refer to what is more precisely described as deterioration in soil physical, chemical and biological properties.

Desertification is the diminution or destruction of the biological potential of land, and can lead ultimately to desert-like conditions. It is an aspect of the widespread deterioration of ecosystems, and has diminished or destroyed the biological potential, i.e., plant and animal production, for multiple purposes at a time when increased productivity is needed to support growing population in quest of development [55].

Chapter 3

THE CASE OF RURAL POVERTY AND ENVIRONMENTAL DEGRADATION IN GHANA: KWAHU SOUTH DISTRICT

DESCRIPTION OF THE STUDY AREA

The district was selected for the study because it lies in the transitional and semi-deciduous forest zone hence very susceptible to degradation. Secondly, it lies in a major food cropping area in Ghana, which is threatened by deforestation, and this has imminent ecological, environmental and socio-economic consequences [56, 25]. The Kwahu Ridge is one of the degraded mountainous areas in the country. The integrity of the original rich forest has been weakened, by heavy logging and wildfires. This renders farming in the area unproductive. Over 30 percent of the forest reserves in the area, particularly, the Worobong and Abisu reserves have been destroyed by wildfires [57].

The Kwahu South District is one of the then fifteen administrative districts in the Eastern Region now comprising Kwahu West, Kwahu East and Kwahu South. It has a total land area of 1,876 km^2 and 750 settlements [58]. It lies between latitude $6^0 30'$ North and 7^0 North and longitude 0^0 30' West and 1^0 West. The district is bounded on the south by Birim North, East Akim and Fanteakwa districts; on the east by the Kwahu North (Afram Plains) District; on the west by the Asante Akim South District and on the north by Sekyere East District. The district lies within three physiographic regions. These include the southern voltaian plateau, which is forest dissected consisting of a series of escarpments, notable among which is the Kwahu scarp rising from

220m to 640m above sea level. The prominent mountains on this scarp are the Odweanoma and Apaku. The other physiographic region part of which happens to fall within the district is the forest-dissected plateau consisting of hills of Birimain rocks which are steep-sided and rise to heights up to 240m. The third is the section of the voltaian basin that stretches into the district. This is a plain rising from 60m to 150m above sea level.

Rainfall in the district can be described as heavy to moderately high with average annual rainfall figures between 1580mm and 1780mm in the forest areas of the Kwahu scarp. The intensity of rainfall decreases as one move toward the voltaian basin. The district experiences double-maxima rainfall pattern with the major period lasting from April to July while the minor one lasts from September to October with a long and pronounced dry season which starts in November and ends in March. The major river in the district is the Afram, a tributary of Volta. It flows through the northern border of the district with the Afram Plains and it is the district boundary. Another is the Pra River that takes its source from the district.

The district also lies within the wet-semi equatorial region. As such, it experiences a double maxima rainfall pattern with average monthly relative humidity ranging between 75 percent and 80 percent during the two rainy seasons. Mean monthly temperature values as high as 30° C and are often recorded between the months of March and April but declines to 26° C in August. The district comes under the influence of two air masses namely the tropical maritime air mass (MT) and the tropical Continental (CT). The tropical maritime air mass hits the district twice a year thereby causing the two rainy seasons. The two occasions are May to August and then September to October. Between the months of November and March, the district is affected by the tropical continental air mass making the area warm and dry.

The district also lies within the semi-deciduous forest zone which belongs to the Antiaris-Chlorophora association. The forest is made of three layers with most of the trees in the upper and middle layers exhibiting deciduous characteristics during the dry season. Relatively little of the virgin forest in the district remains today, due to negative effects of human activities such as timber extraction and farming on the environment. Most areas have been replaced by secondary bush or forest which is easily distinguished from the climatic climax vegetation. The only areas in the district where the ecological balance has not been very much disturbed are the forest reserves.

The study area with an inter censual growth rate of 6 percent [58] and total fertility rate (TFR) of 6.2 [59] is one of the fastest growing areas in Ghana in terms of population. A cursory look at the once evergreen forest of the area in terms of vegetation shows a drastic destruction of the environment due to human activities. It is therefore the aim of this study to examine the relationship between environmental degradation and the level of poverty in the study area.

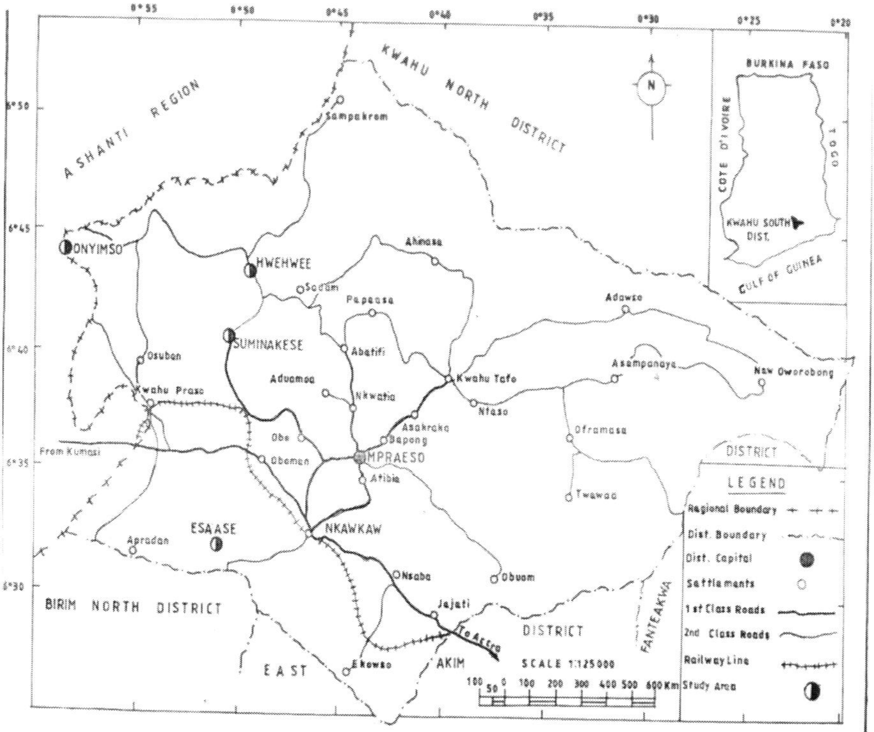

Figure 1. Map of Kwahu South District showing the study area.

THE STUDY POPULATION

The 2000 Population Census put the population of the Eastern region as 2,106,696 with study district which is the Kwahu South district as 217,485 [58]. Furthermore, the total population of the sampled communities is 2,516 [58]. The four communities were Hweehwee, Suminakese, Esaase and Onyimso which were purposively selected for the study. According to the 2000 population and housing census, the adult population for the four sampled communities was 2,358 and the target population for the study was 1,250 which was made up of the adult population (30 years and above). The 2000 population and housing census report indicates that 53 percent of the total population of Ghana were aged 18 years and above. A sample size of 250 respondents representing 20 percent of the adult population was sampled. The twenty percent was based on the adult population size of 1250 which was considered enough to make generalizations for the adult population.

Research Design

A design, according to [60], is a structural perspective of the research that guides the researcher in data collection and analysis. Different types of designs influence different research circumstances. In the case of this research, the design adopted was the survey research design. This involves finding out the persistence and existence of a particular social phenomenon. The survey was cross-sectional since data was collected within a specified period. The instrument for data collection was applied on a sampled population to represent the entire population. Survey design was appropriate for this research because the purpose of the study was to find out the relationship between land degradation and poverty in the Kwahu South District. This involves finding out causes of land degradation and poverty and the relationship between the two hence an analytical method would be suitable in this direction. It was also both quantitative and qualitative in approach because it attempts to establish causal relationships between land degradation and rural poverty. The qualitative method was used to collect data that could not be unearthed using the quantitative approach such as personal experiences and behaviour patterns.

Sample and Sampling Procedure

Multi sampling technique was adopted to arrive at a sample of 250 respondents. The initial process involved cluster sampling which grouped the district into four clusters according to the extent of land degradation as classified by the Forestry Service in the district. Each cluster had between 10 and 15 communities and these were ranked by the District Forestry Service according to their extent of degradation (whether the communities were least degraded, degraded or most degraded). Therefore, the most degraded community was purposively selected from each of the four clusters. It was felt that the four communities could be used to bring out the dynamics between poverty and land degradation in the Kwahu South District because they were homogeneous in terms of levels of degradation. The four communities equally share common population, economic and social characteristics. The selected villages were Hweehwee, Suminakese, Esaase and Onyimso.

Secondly, houses were selected through systematic sampling. The houses in each of the four sampled communities were counted and numbers assigned to them to form the sampling frame. From the list of houses, 250 were selected to form the sample size through the systematic sampling method. A house was randomly selected as a sample and every nth item (third house) on the list was selected to be part of the sample. The number of houses allocated to each community was proportionally done according to the number of houses listed in each community.

In each house, the household head or his/her representative was identified and one household head was sampled randomly until the 250 sample size was arrived at. A head of a household in Ghana, according to the 2000 population and housing census, is the person in the household recognized as such by other household members. This was generally the person responsible for the upkeep and maintenance of the household. The 250 then becomes the sample for the study.

Also, opinion leaders from each community were purposively selected based on their kind of information, expertise and experiences. These leaders were varied as the researcher interviewed religious leaders, assemblymen, chiefs, elders, teachers and people deemed influential by the people in the various communities. The breakdown of the sample sizes for the respondents in the four sampled communities is displayed in Table 1.

Table 1. Distribution of population and sample for the study

Communities	Population size	Adult Population	Sampled Adult Population
Hweehwee	849	450	90
Sumankese	660	350	70
Esaase	472	250	50
Onyimso	377	200	40
Total	2,358	1,250	250

Data Collection Instruments

Both primary and secondary data were used for the study. The primary data was collected from the respondents during the survey conducted in the study area. The survey data were supplemented with data from the district administration and other relevant sources. Documentary sources such as books, journals, articles and other publications provided secondary data. Institutions such as the Ghana Statistical Service and the Kwahu South District Assembly provided valuable information for the study.

Structured interview and questionnaire were the main instruments for data collection. The interview was used to collect data from the illiterate segment of the respondents and literates who could not get time for the self-administered questionnaire because of their busy schedules. The questionnaire was used as a guide for the interview schedules. In such cases, questions in the questionnaire were read out in the local language and the responses recorded. This was of immense benefit to the researcher as doubtful questions were elaborated upon to the complete understanding of respondents. Though this mode of administration was time consuming it elicited valuable information.

The interviews enabled the interviewers to evaluate the respondent's non-verbal behaviour and habits. In the interview process, the researcher and his research assistants made sure that questions were understood properly before the responses to them were recorded. The flexibility of the interview made it an appropriate technique for the exploration of areas where there was little basis for knowing either what question to ask or how to formulate them. It was seen as the most appropriate technique for revealing information about complex and emotionally laden subjects or for probing the sentiments that might underline an expressed opinion. There was also high level of co-operation with personal interviews. Respondents who could answer the questions on the questionnaires were permitted to complete the questionnaires themselves at their convenience.

The questionnaire used for the study consisted of nine sections. Section 'A' collected data on socio-demographic background of respondents; 'B' on eating habits; 'C' on dwelling-related indicators; 'D' on other assets indicators; 'E' on health issues; 'F' on land related issues; 'G' on farming practices; 'H' perception on land degradation and 'I' perception on reduction in output levels.

To ensure that sufficient data was collected for the study observation was used to ascertain some of the activities of the people which were considered to be pertinent to the issue under study such as methods of preservation of farm produce, and the use of forest reserves. The observation also provided data on the physical characteristics of the communities such as types and materials for buildings, roofing, sanitation, social and physical infrastructure and the extent of land degradation. The observation was non-participant.

To further enrich the study, a focus group discussion was also conducted. Discussants were made up of opinion leaders and members from the four communities chosen to obtain in-depth information on perceptions of the people about definitions of poverty and land degradation, their causes and manifestations and how poverty and land degradation could be minimized. The research work was carried out with three research assistants over a three week period.

Pre-testing of the instruments for the study was conducted at Oheneakura, a village outside the four selected communities within the most degraded area to check for validity and reliability a month prior to the actual data collection. It also helped the researcher to acquaint himself with the challenges that are likely to be met on the field during the actual study. Corrections were made to the instruments to cater for the inadequacies observed during the pre-testing.

Methods of Data Analysis

The data collected was statistically analysed by the use of the Statistical Product and Service Solutions (SPSS) version 16 to generate both descriptive and inferential statistics. SPSS generated statistics which included percentages, frequency tables, and means, cross tabulations, chi-square and ANOVA. Cross tabulations and chi-square were most appropriate for the analysis of the data because some data obtained were nominal such as sex, age, occupation and these also constituted independent variables for the study. Also, the chi-square helped to establish relationships among variables such as the relationship between land degradation and output or yield. A five point likert scale which was ordinal was used to rank the perceptions and opinions that were not easy to be quantified. The values assigned to the five point likert scale were: strongly agree = 5, agree = 4, do not agree nor disagree = 3, disagree = 2 and strongly disagree = 1. Qualitative data such as focus group discussions were transcribed and themes identified and used for analysis. Narrations, quotations and pictures were also used to analyse the qualitative data.

Chapter 4

DEMOGRAPHIC CHARACTERISTICS OF RESPONDENTS

This chapter presents the socio-demographic characteristics of the respondents. Among the variables discussed are sex, age, level of education, occupation, marital status, religion and family size. The analysis of the socio-demographic characteristics helped to ascertain the views of the different categories of respondents in the population in terms of poverty and land degradation in the study area.

SEX DISTRIBUTION OF RESPONDENTS

The sex distribution of respondents is necessary because women on all scores are poorer than men as indicated in the Ghana Living Standard survey [61]. Women also are more directly affected when the environment is degraded since they depend on the environment for firewood, water, and farming.

The sex distribution of respondents indicates that 132 of them representing 53 percent were males while 118 (47%) respondents were females. Since information was solicited from household heads, the results demonstrate that most households' heads in the study area were males. The distribution of the data from the study however does not conform to the national, regional and district level statistics from the [34]. At the national level the majority (51%) of the population in Ghana were females whereas 49.0 percent were males [34]. The regional level statistics show that females constituted 50.8 percent and the males constituted 49.2 percent. At the district level the statistics indicate that 50.1 percent of the population were females while 49.9 percent were males. The statistics from the district and regional levels conform to the national trend, however the results of the study does not conform to the national, regional and district level statistics [34]. Therefore, the majority of the respondents being males according to the research are due to the fact that men form majority of household heads in the study area.

AGE DISTRIBUTION

The age distribution of respondents was also necessary as this would help to access the knowledge of respondents regarding the problems of poverty and land degradation. The age distribution would equally inform the study of the age of the respondents. The youthful nature of the population coupled with the high illiteracy rate, lack of access to credit and poor farming practices push the farmers to farm on marginal lands. Continuous cultivation on marginal lands would result in decrease in productivity which would subsequently affect income levels of the people.

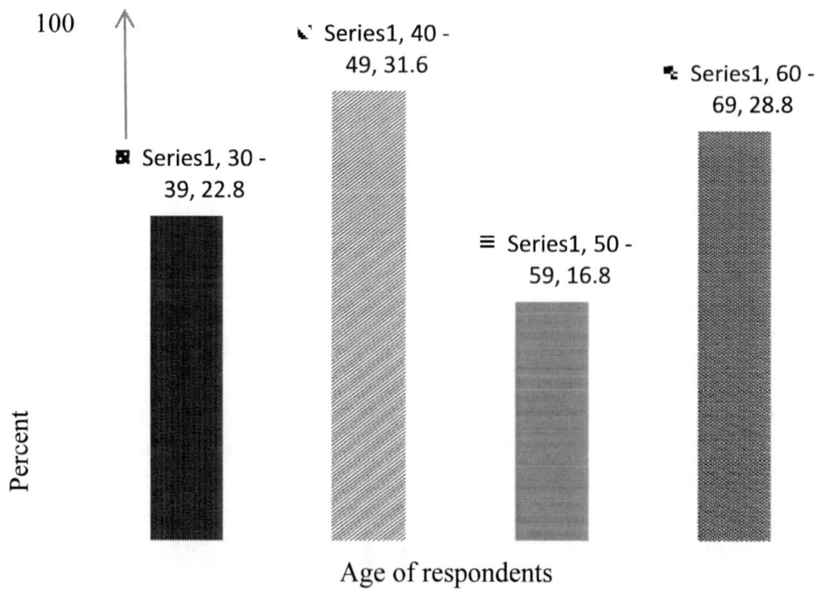

Figure 2. Age distribution of respondents.

Figure 2 shows that 22.8 percent of the respondents were in the age category of 30-39, while 32 percent were in the 40-49 age categories. About 28.8 percent were in the older age of 60-69. About 54 percent of the respondents were less than 50 years of age. This is comparable to the regional statistics in that the active age of the population 15-64 for males stood at 51.6 percent while that of females was 53.6 percent. This implies that at the regional level there are more females in the active population than males.

However, a high percentage of the population is youthful as indicated by the 2000 population where the active age group (15-64) years is made up of 49.3 percent males and 50.7 percent females. Since the predominant economic activity in the study area is agriculture and 78 percent [34] are in the youthful age of the population within the active working population, it means large tracts of land would be put under cultivation. The continuous cultivation of arable land will results in loss of soil fertility and deforestation, which in turn leads to reduction in crop yields.

Level of Education

The educational status of the respondents follows the normal trend in Ghana where the majority of the rural populace are either illiterates or have only basic education. The low levels of education of respondents might influence how they accept improved methods of farming, such as the use of genetically improved seeds and fertiliser. The non-acceptance of modern methods of farming could result in the reduction of output levels of farmers, which would subsequently lead to the reduction in income level. Thirty-one percent of the respondents had no formal education. This is below the national illiteracy rate of 37 percent in the 2000 population census report. The majority (57%) of the respondents had basic education as the highest level of educational attainment. Only nine (3.6%) of the respondents have had some form of tertiary education in the form of Training College, Polytechnic or University education. About three percent of the respondents reported to have had other forms of education such as vocational and technical.

From the statistics presented, 88% of the respondents had no education or at most the basic level as the highest level of education attained. The general low level of education may explain why most of the respondents engage in agriculture as the primary occupation. Table 2 shows the relationship between sex of respondents and education. From the cross tabulation in Table 2, the majority of females (63.4%) did not have any education while the males were

36.6 percent. The lack of education on the part of females in the study area can have effects on land degradation and poverty.

The number of females without any level of education would have little or no knowledge as regards to modern farming practises such as fertilizer application. Therefore, this could lead to over cultivation and over intensification of the land leading to land degradation.

Table 2. Sex by education of respondents

	Level of education							
	None		Basic		Secondary and above		Total	
Sex	F	%	F	%	F	%	F	%
Sex	F	%	F	%	F	%	F	%
Male	26	36.6	72	50.7	34	91.9	132	52.8
Female 45	63.4	70	49.3	3	8.1	118	47.2	
Total	71	100.0	142	100.0	37	100.0	250	100.0

Table 2 shows the relationship between sex of respondents and educational level. However, a chi square test that was performed did not produce significant result as the results indicated a chi-square value of 30.397. This means that the educational level of respondents in the study area is independent of respondents' sex. The fact that 63.4 percent of females do not have any level of education has a negative consequence on land use. The high level of illiteracy among females confirms the notion in most traditional societies that the place of a woman is the kitchen and other social factors such as early marriage and the general attitude of the people is preference for boy's education to girls.

The level of education could have a positive impact on land in terms of its usage and management and good farming practices to prevent land degradation. However, Table 2 shows an interesting scenario where, out of the total number of respondents no female had tertiary education while 3.6 percent of males had tertiary education.

Occupation of Respondents

In terms of occupation, 78.4 percent of the respondents were self-employed in agriculture and agricultural related activities as shown in Figure

3. This follows the national trend in the country where over 60 percent of Ghanaians depend on agriculture as the main source of livelihood. About 2.8 percent of the respondents were in non-agricultural employment with 8.8 percent in other activities such as carpentry, masonry among others. The low educational background of the people could explain why only 10 percent were in salaried employment.

The occupational status of the respondents is a clear indication that the people depend on the land for their livelihood. The majority engaging in agriculture means more land would be needed for cultivation and with the low educational levels, appropriate knowledge and methods of modern farming practices might be lacking. This could subsequently lead to land degradation.

The occupational status of the respondents also gives the idea that there are no alternative jobs available to the people and that their major sources of income would be limited to only agriculture and its related activities. Figure 3 shows the occupational status of the respondents. The occupational categories of respondents are presented in Table 3. More males (40%) were engaged in self-employed agriculture whilst females constituted 38.4 percent. Non-agricultural self-employment recorded 5.6 percent of males and 6 percent of self-employment such as petty trading, artisan etc.

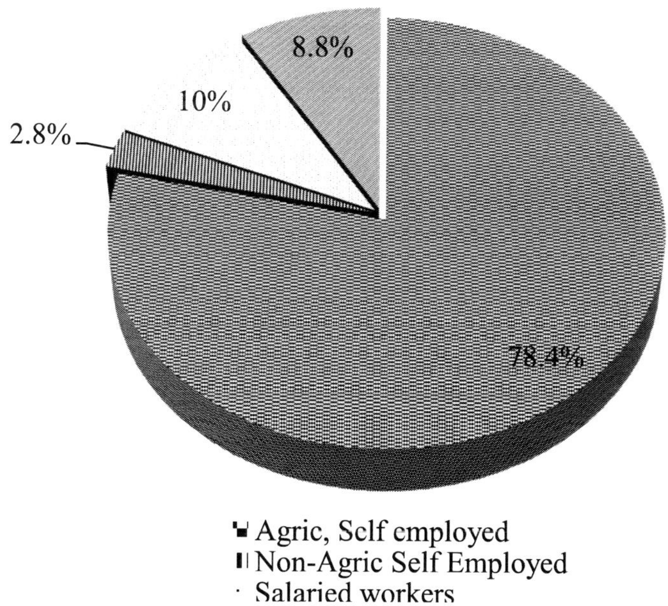

Figure 3. Occupation of respondents.

Table 3. Sex by primary occupation

Primary occupation								
Sex	Agricultural self employed		Non agricultural self employed		Salaried		Total	
	Freq	%	Freq	%	Freq	%	Freq	%
Male	100	40.0	14	5.6	18	7.2	132	52.8
Female	96	38.4	15	6.0	7	2.8	118	47.2
Total	196	78.4	29	11.6	25	10.0	250	100.0

Salaried employment recorded 7.2 percent for male respondents and 2.8 percent of female respondents. A chi-square calculated value of 4.19 was observed compared with the statistical value of 5.991 using a significance value of 0.05 on the association between sex and occupation showed that sex does not influence occupation of respondents.

Therefore, the number of agricultural self-employed, non-agricultural self-employed or salaried workers does not depend on whether the respondent is male or female (Table 3). The relationship between agriculture and rural poverty is given credence by [40] who attributed poverty to the type of occupational structure of most rural communities in Africa. The biological nature of agricultural production means that it could not be increased or reduced as rapidly as manufactured goods. This made it impossible to adjust production to changes in consumer demand because of long gestation periods. Also, the volume of production varies unpredictably from year to year according to the weather. This made agricultural income unpredictable and unstable. Agriculture is vulnerable to parasites.

Marital Status of Respondents

Marital status of people could determine the fertility rate of a country and the average family size within that country. The data indicates that where the family size is large there will be more mouths to be fed. Thus more pressure would be put on the land and this can lead to land degradation if sustainable framing practices are not adopted. If the land is degraded families would be dependent on marginal lands or degraded lands which could lead to low yields with more mouths to feed and the resultant consequence is poverty.

About 16.4 percent of the respondents, reported being single. The majority (66.8%) were married. Another 17.2 percent of respondents were divorced, separated or widowed.

Slightly more than 44% of the respondents had family sizes above five members. It is also a common phenomenon in Ghana given that most families have communal living arrangements. This incidence of large family sizes in terms of the number of children could impact negatively on such variables as education, health care, and feeding. Some respondents (28.8%) reported their readiness to give birth to additional children. This has a lot of resource implications in terms of income, standard of living and access to social facilities such as education, health care and sanitation among others.

FAMILY SIZE

Large family sizes are a common feature in the study area as the average family size of five members per household was recorded in the study area. The data indicated that 67.8 percent of the respondents had more than five members in a household. According to the GLSS4 categorisation any household with more than five members in Ghana can be classified as a large household. Table 4 displays the family sizes of respondents in the study area.

Table 4. Family sizes of respondents

Family sizes	Frequency	Percentage
1-4	98	32.2
5 above	152	67.8
Total	250	100

The motivation for people to have large family sizes was supported by a respondent at Hweehwee as follows.

> "Over here most of us see children as security or insurance against old age. We are unable to save financially for our respective old ages, and the children are our only source of hope. Our future happiness or otherwise depends largely on the number of children that we have. Children are also seen as gifts from God or the gods hence, we need not to temper with any unwanted pregnancy. I did not want to offend the gods so if I had become pregnant twenty times I would have had twenty children.

> Some men over here are desirous of having ten children in order to qualify for the traditional "badu guan" that is a sheep given to a husband by his wife's family for adding ten more human beings to their family. Also, because we are Akans and therefore, practise matrilineal system of descent and inheritance, maternal uncles or family heads take pride in the largeness of their lineages, so they would stop at nothing but encourage their in-laws to increase the family sizes. People are also proud of the number of children that they have, hence large family sizes. Another motivation for us to have large families is to use them as source of labour on our farms. We also lack access to improved medical care so there is the fear and possibility of some of the children dying before reaching adulthood so we need to have more children such that in the event of death we still have some children".

The large family sizes have resulted in children not staying with their biological parents (Table 4). Fifty-five percent of the respondents indicated that they had children who lived with family members, friends or benefactors. In most cases, the rationale behind sending children to live with others was due to the fact that the real biological parents themselves were often over burdened with so many problems to the extent that they were unable to provide adequately for these children. This was a result of the fact that their poverty situation was very endemic. Some parents also wanted their children to go to some of these relations in the urban areas so as to receive better and quality education, which were not available in rural.

Chapter 5

MANIFESTATIONS OF POVERTY IN RURAL AREAS OF GHANA

GENERAL OVERVIEW

This chapter examines the manifestations and causes of poverty. This is done to help answer one of the research questions of the study. The first part of the chapter therefore, presents the manifestations of poverty whereas the second part examines the causes of poverty in the study area.

Absolute poverty means lack of food, cash and any assets. It is a 'condition' of life so degraded by disease, illiteracy, malnutrition and squalor as to deny its victims basic human necessities. It thus means a condition in which the possibility to obtain the basic needs of life or where deprivation is so severe that the basic needs of life can scarcely be met at the minimum level of survival. In other words, absolute poverty is the state of existence in which the overall needs of the individual are not satisfied due to lack of enough purchasing power or means for self-provisioning.

Poverty manifested itself in various ways in the study area. Among them are inadequate access to basic necessities of life (i.e., food, clothing and shelter), high level of illiteracy, poor quality of education, poor housing and poor sanitation and environmental degradation.

Inadequate Food as Manifestation of Poverty

Absolute poverty can be described as a situation in which people barely exist where the next meal may literally be a matter of life or death as the cumulative effects of malnutrition and starvation [39]. Existence becomes a matter of struggle because people who find themselves in this kind of poverty do not have any hope of getting the next meal and therefore below or beneath this level of existence is only death.

An important manifestation of poverty in the area was the fact that respondents either borrowed food, relied on relatives or friends, limit intake or reduce number of meals per day. Table 4 explains the respondents' views on inadequacy of food in study area.

Table 5. Food related indicators of poverty

Indicators	Never		Yes		Total	
	Freq	Percent	Freq	Percent	Freq	Percent
Borrowed food/money per week	150	60	100	40	250	100
Rely on relatives or friends within the week	120	48	130	52	250	100
Limit intake of food per week	145	58	105	42	250	100
Reduced number of meals per week	140	56	110	44	250	100

Forty percent of the respondents contended that they borrowed either food or money before they could meet their food needs. In addition, 52 percent of the respondents (Table 5) reported their reliance on help from either a relative or friend outside their households before they were able to provide the food needs for themselves and their children. Fifty eight percent of the respondents reported that they had never limited their food intake to ensure sufficiency for their children, while 42 percent indicated that sometimes, they had to limit their own intake to ensure that their children get enough to eat. Families also skipped some meals they took a day in order to satisfy the needs of their children. Forty-four percent of the respondents reported that on certain days of some weeks, they had to reduce the number of meals eaten so as to satisfy the children and save some for the next day. This was in line with [39] definition of absolute poverty described as a situation in which people barely exist, where the next meal may literally be a matter of life or death as a cumulative effect of malnutrition and starvation.

The poor quality of food intake could result in many health problems particularly increased mortality among the people. According to the municipal director of health services malnutrition is a major health problem within the municipality and that it accounts for about 20 percent of infant/child mortality in the municipality. This is supported by the data from the district hospital which revealed high outpatient department attendance and increased incidence of mortality over the period 2006-2008. This is shown in Table 6.

This problem was aggravated by the inadequacy of health facilities in the district, which prevented the people from accessing medical treatment (Table 6). The nearest health facility, the Nkawkaw Holy Family Hospital, is about 10km away from each of the sampled communities.

Table 6. Distribution of OPD, Inpatients and mortality cases at the Holy Family Hospital Nkawkaw

Year	Out Patient Dept	In-patients	Mortality
2006	115623	8649	351
2007	120899	10823	379
2008	121855	11441	392

Housing as a Manifestation of Poverty

The dwelling places of the people also showed the extent of poverty in the area. Even though some of the respondents owned their own houses and housing units which constitute 20 percent of the number of respondents, 62 percent of them did not own any form of house or housing unit with 18 percent renting. About 29 percent of the respondents occupied only one room with their dependents whereas another 43 percent used two rooms with their families. This revelation is contrary to the regional and district averages revealed in the Ghana Statistical Service, 2000 population census report. According to the census results the average number of household members owning their dwelling unit in the region is 58.2 percent, ranging from a high of 74.2 percent in the Afram Plains, 70.9 percent in Birim North, Kwahu South

52.2 percent to a low of 48.4 percent in Asuogyaman, and 33.9 percent in New Juaben.

The results of the house ownership (38%) from this study as compared to the figure from the population census (52.2%) indicate that the majority of the respondents from the study area did not own houses. This is an indication that the study has revealed a contrary position as far as the ownership of houses in the study areas is concerned. This could be due to the nature of the sampled areas being most degraded and poor and therefore such respondents did not own houses.

However, the rural stock of houses accounts for 73.4 percent and the urban, 26.6 percent in the district. This implies that though there are more houses available in the rural areas most rural dwellers do not own them. This is a clear indication of poverty.

Also the nature of housing further demonstrates the poor situation of the people in not only the study area but also in the whole district. The 2000 population census revealed 46.5 percent of the houses were constructed with mud or bricks or earth. Soil erosion was evident in the environment resulting in big gullies in the villages as shown in Plate 1. This left some houses hanging which could collapse whenever there was any strong wind or rainstorm.

Plate 1. A hanging building at Suminakese due to erosion.

Plate 2. A residential place at Hweehwee.

The materials used for the buildings were mainly mud with raffia palm as roofing sheets. Some of the houses are so dilapidated that they pose a threat to human life. A few of these buildings, even though built with mud, are cemented and roofed with corrugated iron sheets. The state of housing units in the area is shown by Plate 2.

Social Indicators of Poverty

The high incidence of adult illiteracy and high school dropout rates are another manifestation of poverty in the area. About 31.2 percent of the respondents have had no form of formal education, while a significant proportion of 57.2 percent only managed to attain either Junior High School or Middle school education as the highest level of educational attainment. On the question of the highest level of education respondents would like their children to attain, as much as 27.6 percent expressed their happiness with only basic or secondary school education, some did not express willingness for their children to go to school at all.

The fact is, due to poverty, parents were unable to provide some basic school materials in the form of school uniforms, pencils, pens and exercise books for their children. Thus, it was not as if they were unwilling to either send their children to school or for them to attain high levels of education but they could not afford. Some parents due to their level of education had not as yet realized the value of education hence; they were unwilling to send their children to school. Using torn and worn out school uniforms for classes and using timber boards and blocks as furniture in schools were common in the study area. Children whose parents were unable to provide their educational material needs became disinterested in schooling, and hence the high school dropout rate in the area.

The situation of education revealed in the research was corroborated with the district and regional averages revealed by the 2000 population census report on the state of education in the Eastern region. According to the 2000 population census report, the educational attainment at the various levels of education in the region is not satisfactory. A large proportion of the school population do not proceed beyond the primary level, with those at the Junior High and Senior High levels fairing is even poorer.

In the region, about 22.5 percent of the pupils terminate their education at the primary level, made up of almost the same proportion of males 25.8 percent and females 25.2 percent. At the JHS level, 30.1 percent complete the level, made up of a third 33.4 percent of the males and 26.8 percent of the females, while at the senior secondary school level, 5.8 percent, complete the level.

A significant number of pupils and students drop out of school at the various levels, mainly because of lack of adequate facilities at the next level. The number of Junior High Schools is just about half the number of primary schools, while there are only 74 Senior Secondary Schools and less than five Vocational and Technical (VOTEC) institutes, and one Polytechnic in the region to absorb all the JSS leavers. This point to the urgent need for more physical infrastructural facilities, and staff for the schools in the district of the region.

In terms of educational facilities, all the communities had access to at least a primary school. However, the state of most of the school buildings leaves much to be desired. Plate 3 shows the classroom block of Suminakese Local Authority Primary School which was an ample demonstration of the dilapidated nature of the school building. For furniture most schools rely on timber boards and block in the study area.

The inadequacy of social amenities in the area is another manifestation of poverty. Most of the communities are saddled with problems of poor sanitation, lack of toilet facilities, inaccessible roads, and unsafe drinking water supply. Communities did not have even a pit latrine hence; indiscriminate defecation was the main mode of excretion. Commonly reported health problems included malaria fever, typhoid fever, worm infestations, and diarrhea which are related to poor sanitation. The plight of the people is worsened by the lack of health facilities.

Apart from Hweehwee which has a clinic, with the rest of the communities, the nearest health facility is the Holy Family Hospital at Nkawkaw with the closest settlement being 30km away with unmotorable feeder roads. The nature of the roads linking some of the villages to the major towns was such that a 10km road could take over one hour to cover. The roads were full of potholes and gullies making transportation very difficult. Farm produce especially perishable food items go bad due to transportation difficulties. Farmers are therefore forced to sell their goods at give-away prices which make it impossible for them to breakeven.

On the type of energy used for cooking, for instance, most respondents reported their dependence on firewood and charcoal. About 80 percent of the respondents used firewood as a source of energy for cooking whereas 15.5 percent of them resorted to the use of charcoal. The use of fire wood and charcoal constituted 95.4 percent of energy used for cooking in the study area. This has serious implications for land degradation in the form of deforestation and soil erosion. Perhaps this could be explained to mean that, they were not in the position to afford such energy sources as electricity, gas or kerosene.

Plate 3. Suminakese Local Authority Primary School.

Asked to comment on the relationship between the District Assembly and the central government on one hand, and the communities on the other, respondents were not happy with the relationship. They were very bitter with the Assembly and the central government in terms of the sharing of the national cake. The picture in Plate 4 shows the extent of neglect of the communities in terms of provision of infrastructure. The traders engage in their economic activities in the scorching sun whereas the Assembly was interested in the collection of market tolls for its urban developmental projects. The situation regarding social amenities points to the fact that the quality of life of the people in the study area is low.

The low standard of living of the people has compounded the plight of the people. They were unable to finance any locally initiated projects themselves to improve upon the social infrastructure situation. Basic necessities of life (i.e., food, shelter and clothing) were left unmet. Even though communal labour was organised periodically to see to the provision of some of their needs, participation in these activities by the youth is usually low. Apart from a few communities such as Hweehwee and Suminakese which enjoyed potable drinking water from boreholes, most of the settlements within the study area depended on streams and hand dug wells as the main source of drinking water.

Plate 4. A market scene at Hweehwee.

Table 7. Ownership of household assets

Household assets	Responses					
	Yes		No		Total	
	Freq	Percent	Freq	Percent	Freq	Percent
Car	6	2.4	244	97.6	250	100
Motor cycle	4	1.6	246	98.4	250	100
Bicycle	46	18.4	204	81.6	250	100
Television	55	22	195	78	250	100
Stereo, video, deck	35	14	215	86	250	100
Radio/ tape	187	74	63	25.2	250	100
Fridge	43	17.2	207	82.8	250	100
Washing machine	1	0.4	249	99.6	250	100
Sewing machine	90	36	160	64	250	100
Electric fan	40	16	210	84	250	100

HOUSEHOLD ASSET AS AN INDICATOR OF POVERTY

The indication of poverty in the study area also manifests itself in most of the people without asset holdings. About 25.2 percent of the respondents did not possess household items such as radio, television, tape recorder or furniture. Out of the 250 respondents, only 9.5 percent possessed some form of productive assets while 90.5 percent did not have any productive asset. This made it highly difficult for the people to generate any income coupled with the fluctuating farm incomes. Table 7 depicts the household's assets holdings of the respondents.

Even though the people in the area are farmers, there is low capital formation and this is shown by the number of people who can save either in the form of grains or rearing of animal and poultry (Table 7). It is not surprising that the people could not afford even fertilizers on their farms that can increase the productivity of their farms to move them from largely subsistence cultivation to some form of commercial farming.

ENVIRONMENTAL INDICATORS OF POVERTY

To establish environmental factors that can influence reduction in output levels hence income levels (poverty), respondents were asked to rank in order of priority various factors that could affect output levels. Table 8 shows the ranking as indicated by respondents. In arriving at the weighted means for the data, values were placed on a five point likert scale of 1-5 where 1 = very good; 2 = good; 3 = not good or poor; 4 = poor and 5 = very poor. The weighted means were calculated by multiplying each frequency by its weight (1, 2, 3, 4 and 5) and take their sum total. This sum total was then divided by the total number of respondents (250).

The high level of poverty in the area is the results of various factors as shown in Table 8. From the table, factors such as bush fires, loss of soil fertility, insufficient rainfall and cutting of trees for timber with their respective means of 1.61, 1.68, 1.77 and 1.67 gives credence to the assertion that environmental factors could result in low output hence poverty.

Table 8. Environmental factors responsible for reduction in output levels hence income levels (poverty) of respondents

Practices	Responses					Mean	Std.D
	SA	Ag	NA	DA	S D		
Bush fires	135	95	8	6	6	1.61	0.892
Loss of soil fertility	134	89	10	8	9	1.68	1.065
Insufficient rainfall	114	106	12	10	8	1.77	1.075
Cutting of trees for timber	115	95	14	10	6	1.67	0.970
Destruction of crops by wind	93	125	10	15	7	1.87	1.055
Cutting of trees for charcoal burning	104	110	15	10	9	1.82	1.082
Cultivation of a land continuously for more than five years	98	114	16	13	9	1.88	1.050
Destruction of crops by insects	85	127	18	10	10	1.93	1.056
Prolonged drought	125	71	27	17	10	1.82	1.082
Inability to adopt modern farm practices	72	129	30	13	6	2.01	1.086
Small farm size	76	117	40	15	2	2.00	1.084
Soil erosion	67	136	27	10	10	2.00	1.084
Excessive rainfall	97	84	40	16	13	2.06	1.092

Farmers, due to poverty, cultivate marginal lands, which should have been allowed to fallow under normal circumstances. Yields per acre of such marginal lands are therefore very minimal. Cultivation of marginal lands coupled with small farm sizes due to land fragmentation makes it very difficult for the people to get something for subsistence let alone saving to generate capital for investment, which would move them out of the poverty trap (table 8). The cultivation of marginal lands has also resulted in land degradation in the form of deforestation. The only river that takes its source from the district, river Pra with its tributaries, easily dries up during the dry season due to human activities and pressure. This makes farmers depend solely on rain fed farming which is also erratic particularly in the lean season. The net result is that these human activities impact negatively on the environment hence exacerbating the poverty situation in the study area. Firewood which is the main source of energy for the people also exerts considerable effect on the forest.

The unwillingness and inability of the people to adopt new technologies as sources of energy for cooking has resulted in firewood being the only source of energy. This has resulted in deforestation. Figure 4 shows the percentage usage of sources of energy in the study area.

The once tropical rainforest vegetation of the area has turned into shrubs and *Chromolena odorata* ("Acheampong") resulting in erratic and poor rainfall patterns in the area. The environmental factors that cause poverty in the study area are shown in Table 9.

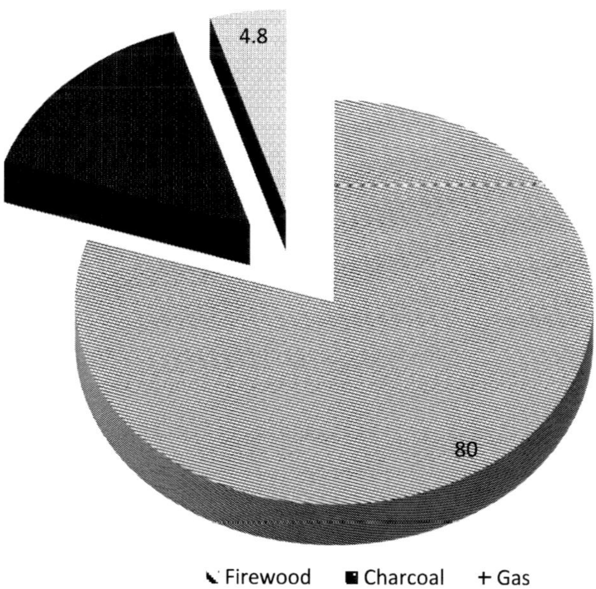

Figure 4. Sources of energy for cooking.

Table 9. Evidence of environmental degradation

Item	Very Good	Good	NG/P	Poor	V. Poor	Wt. Mean	SD
Rainfall pattern							
10 years ago	190	50	5	5	0	1.30	0.609
5years ago	10	197	6	33	4	2.29	0.806
Current	2	42	5	110	91	3.98	1.064
Vegetation of the area							
10years ago	181	59	4	5	1	1.34	0.653
5years ago	16	199	5	30	0	2.19	0.726
Current	3	41	6	111	89	3.96	1.071
Soil fertility							
10years ago	191	41	4	6	0	1.30	0.622
5years ago	12	203	42	2	2.22	0.742	0.622
Current	3	32	6	129	80	4.00	0.983
Yield per acre							
10years ago	186	50	5	7	2	1.35	0.731
5years ago	6	184	3	44	3	2.33	0.882
Current	6	40	4	110	90	3.95	1.111
Farm size							
10years ago	132	79	3	32	4	1.78	1.075
5years ago	16	189	7	33	5	2.28	0.848
Current	29	54	8	94	65	3.44	1.379
Income level							
10years ago	146	64	4	34	2	1.74	1.070
5years ago	8	178	6	51	7	2.48	0.948
Current	20	38	5	89	98	3.82	1.310

In arriving at the weighted means for the data values were placed on a five point likert scale of 1-5 where 1 = very good; 2 = good; 3 = not good or poor; 4 = poor and 5 = very poor. The weighted means were calculated by multiplying each frequency by its weight (1, 2, 3, 4 and5) and take their sum total. This sum total is then divided by the total number of respondents (250).

To indicate the effects of variables such as rainfall, vegetative cover, soil fertility, and yield per acre and income level on land degradation, the opinions of respondents regarding the effects of these factors on land degradation were sampled and this was done using the weighted mean method and likert scale ranking of very good, good, not applicable, poor, and very poor

The weighted means of the responses to patterns and trends of certain indicators over the past ten years is a clear indication of worsening environmental conditions of the study area. The weighted means and the time frame were negatively related. For instance the weighted mean of rainfall pattern 10 years ago of 1.30; 5 years ago of 2.29; and current rainfall pattern of 3.98 shows a deteriorating trend over the years. The same goes for vegetative cover and loss of soil fertility. The indicators show that the environmental conditions over the past 10 years were better than the current environmental conditions.

Similarly, a yield per acre ten years ago with mean of 1.35 as against 3.95 currently, suggests a reduction in income levels of respondents hence an increase in the poverty situation. This was collaborated by the means of income levels ten years ago of 1.72 as against current income level of 3.82. The statistics as presented in Table 9 suggests a positive relationship between land degradation and poverty. For instance, higher soil quality, signifies higher output levels of the people as the years go by, as compared to current times.

Chapter 6

CAUSES OF POVERTY

The study also examined the specific causes of poverty in the study area. This was aimed at isolating general causes of poverty in the literature from what pertains specifically to the Kwahu South District.

FAMILY SIZE AND RURAL POVERTY

High population growth rate and large family sizes are common in most rural communities in Ghana. This is in line with [40] view that population growth has a lot of implications which may result in poverty. Population pressure also forces people to farm on land of lower quality. High population growth rate may result in increased dependency ratios, division of assets between several children, land fragmentation and diminishing returns to labour in the absence of any new technologies or farming systems.

Large family size was a common feature in the study area as the average family size of five members per household was recorded. The data indicated that 60.8 percent of the respondents had more than five members in a household, slightly more than the regional average household size of 4.6 [58]. The 283,461 houses in the district constitutes 13.0 percent of the total stock of houses in Ghana and represent 51.4 percent increase over the 1984 stock of houses in the region. The rural areas' share of the regional housing stock is 73.4 percent, which is higher than the national average of 65.9 percent [58]. However, according to the GLSS4 categorisation any household with more than five members in Ghana can be classified as a large household. Therefore, the results as indicated by respondents confirm the data from the 2000

Population and Housing Census report as the average household size was 4.6 which can be considered as large.

The large family sizes have resulted in most families in rural communities becoming impoverished. This emanates from the fact that consumption levels of such families were higher and per capita income lower. Such families would lag behind since they end up spending every bit of their income on feeding and management of their respective homes. Dependency ratio was very high, increasing the burden of the few economically active people in the area. As a result of this, there was low level of savings and therefore no investment to push them out of poverty trap. To further ascertain this relationship the researcher tested the hypothesis that the larger the family size, the greater the reduction in food intake.

Table 10 shows the relationship between family size and intake of food in the study area.

The cross tabulation of family size and reduction in food intake (Table 10) gives a calculated χ^2 value of 2.196 which was lower than the critical value of 5.991. This means that there is no association between family size and reduction of food intake. This observation was in line with the views of the respondents through the focus group discussion that large family size in itself may not lead to poverty.

They claimed that a large family size was a potential source of labour for their economic activities especially farming. This disagrees with [59] which states that countries in Africa frequently assign low priority to agricultural growth and rural development because; they view agriculture as a declining sector as population increases. It was the view of the respondents that household size does not matter much. Rather, it is the unavailability of resources such as rich farmland or alternative job opportunities that may result in poverty.

Table 10. Family size and reduction in food intake

Family size	Reduction in Food Intake						χ^2	Df	Sig. Level
	Yes		No		Total				
	N	%	N	%	N	%			
1-4	60	36.6	38	44.7	98	39.2	2.196	2	0.05
5+	104	63.4	48	55.3	142	60.8			
Total	164	100	86	100.0	250	100.0			

Large family sizes in the study area emanate from adherence to cultural practices, customs and traditional beliefs. This was justified on the grounds that some of the children due to high infant mortality would die. This may seem to be irrational at this age of modern medical science and improvement in the health delivery system but their fears were justified on the grounds of inadequacy of health facilities, lack of potable water and poor sanitation, exposing both mothers and children to health hazards. A large family size was also justified by the people on the belief that one of the children could be successful or rich in future. Children were also seen as an insurance against old-age. For parents to get someone to fall on in times of need (especially during old age and sickness) they decide to have more children so that each of them would contribute to their up keep when the need arises. The desire of some men in the area to collect the traditional "badu guan" (A reward of a sheep for giving birth to the tenth child) would make them stop at nothing until they receive the ultimate prize.

Men see the number of children they have as a source of pride and they boast about the number of children they have. Some rural dwellers also see children as gifts from the gods so they should not offend their ancestors and gods by using contraceptives or resorting to abortion. The Christians among them also upheld the biblical injunction to "give birth and multiply like the sand at the seashore" so they were fulfilling the wish of God by producing more babies.

The aforementioned factors which naturally would contribute to large family sizes perpetuate rural poverty. A greater part or almost all of what is produced by the family is consumed without selling some to generate income to meet other needs especially health and educational requirements. Because of the large number of children, parents were unable to give their wards any appreciable level of formal education which might lead to securing alternative forms of employment other than farming. It was, therefore, not surprising that poverty was prevalent at the study area. Large family sizes most often result in farmers keeping or storing almost all of their farm produce for home consumption. Small farm sizes coupled with low level of technology did not permit large-scale production. Production was therefore, geared towards home consumption. What was produced was even unable to feed the members of the family let alone selling some to generate income to meet other needs of the family. Farmers even purchased food to supplement what they were able to produce.

Asked whether people bought food to supplement what was produced personally, as many as 85.2 percent of the respondents said yes, while only

14.8 percent said they did not purchase food to supplement what they produce themselves. Also, most farmers stored a greater part of their farm produce for home consumption. As much as 40 percent of the respondents stored all their farm produce for home consumption while 35 percent stored three-quarters of what they produced. The inability of large families to provide the needs of the family members may emanate from the fact that, the family heads themselves did not receive any appreciable level of formal education. They were, therefore, not in gainful employment which could generate the needed income to meet the needs of their families. Formal education which is one of the surest means to escape from poverty trap is even denied the children of such families and the children would become even poorer than their parents when they grow up. Most of these ill-educated children would end up in lowly placed jobs which go with low remuneration. Therefore, it would not be easy for them to move out of the poverty bracket.

EDUCATION AND RURAL POVERTY

Another major cause of rural poverty, as observed from the data, was the problem with low educational level or lack of educational opportunities. In the area under study, it was revealed that education is not so much priced or prioritized [60] explains the relationship between education and poverty.

> Limitations to knowledge and skills due to illiteracy result in the inability to adopt new technologies or adjust the farming system, hence, low output and low income by the rural folks which eventually land them into poverty [60].

[61] Asserts that there is some basis for attributing the poverty of individuals to educational deficiencies. That is, those with low level of education are mostly affected by poverty. To ascertain the relationship between education and poverty, respondents were asked how educational attainment determines primary occupation. Table 11 depicts the relationship between educational status and primary occupation of respondents.

As many as 92.4 percent of those who engage in agriculture have at most basic education as the highest level of educational attainment. This means they do not have any regular and reliable source of income. This gives credence to the assertion that the higher one's level of education, the higher his/her income.

Table 11. Educational status and primary occupation

Level of Education	Agric		Non Agric And Salaried					Total	
	N	%	N	%	χ^2	Df	P	N	%
None	63	32.1	7	13.7	57.1	4	0.00	70	28.4
Basic	118	60.3	24	47.0				142	7.5
Secondary	11	5.6	9	17.6				20	8.1
Tertiary	1	0.5	7	13.7				8	3.2
Others	3	1.5	4	7.8				7	2.8
Total	196	100	51	100				247	100.0

The higher one's educational level, the higher his or her chances of being in a well-paid and good employment. The educational status of the inhabitants of the study area is not satisfactory. Only 11.6 percent of the respondents had accessed secondary education and above. It was therefore, not surprising that only 10 percent of the respondents were salaried workers.

None of the respondents with no level of education was in salaried employment, whereas six out of the eight respondents who had tertiary education (Table 11) were in salaried employment. This suggests a positive relationship between level of education and the type of employment one does. We cannot but agree with the [62] that illiteracy breeds poverty based on the data analysis. The World Bank attributes poverty in Sub-Saharan Africa to low endowment to human capital as a result of inadequate access to education [63] also collaborated the relationship between illiteracy and poverty by asserting that;

> Among the poorest 10 percent of Ghanaians, 79.5 percent live in households headed by someone with no education at all; the corresponding figure for the poorest 30 percent is 69.8 percent. Almost all the rest live in households where the head has only a primary or middle school education (p. 49)

Furthermore, lack of formal education did not allow the people to engage in any commercial production. Even though the respondents were aware of the importance of fertilizer in their farming activities some were disinterested in its use because to them, the nutritional value of farm produce from fertilizer application were of lower quality. It was therefore not surprising that only 5.5 percent of the respondents had used fertilizer in their farming activities one time or the other. According to one of the respondents as captured during the field interview:

Because we are not educated we are unable to access paid employment. The only alternative is to till the land which serves as our only source of income. I have cultivated the same piece of land over the years hence the low output levels. We do not engage in any modern/ scientific methods of farming due to illiteracy. Why should I plant new variety of maize which is not tasty as compared to the traditional maize. Crops planted with fertilizer and other chemicals are also not tasty. I prefer quality to quantity so there is no way that I would accept anything besides what I know.

This problem can however be attributed to the use of mainly simple and obsolete farm tools such as hoes and cutlasses coupled with the non-use of chemicals in the form of fertilizers and manures to enrich the soil. The result is that these farmers end up harvesting very little produce.

ACCESS TO LAND/FARM SIZE AS A CAUSE OF RURAL POVERTY

The difficulties associated with land acquisition, the land tenure system and the small size of most rural farms coupled with the use of rudimentary agricultural techniques and practices were some of the main causes of poverty in the study area. For instance, in a response to a question on how respondents acquired their farm lands and whether they have had any problem in their efforts at acquiring them, only 9 (3.6%) of the respondents said they owned land. The remaining claimed that they rented, borrowed, or engaged in "abunu/abusa" system. Thirty-seven percent of the respondents reported that they had encountered some problems or difficulties in the acquisition of their farmlands. There were often conflicts between the original landowners and the caretakers, while the sizes of most of these farms were often too small. Again, many of these farms were always cultivated on subsistence basis.

FARM SIZE AS A CAUSE OF POVERTY

Scarcity of land resulting in small farm sizes as a cause of poverty has been given a lot of prominence by researchers in many rural poverty studies [64]. Poverty in Africa is caused by corruption and poor governance, poor land utilisation and land tenure system, civil wars and unending political conflicts, poor infrastructure, diseases and poor health facilities [65]. This issue of scarcity of land leading to small farm sizes has been found to be one of the factors responsible for poverty in the study area.

Output levels depend to a large degree on the farm size. The smaller the farm size the lower the output and as farm sizes increase then output also should increases (Table 12). The small farm sizes suggest that output levels are low. This is supported by [66] who attribute rural poverty to subsistence farming and the inability of the poorer rural areas to attract national investment despite their greater need.

The output levels of respondents emphasize the relationship between farm size and output of crops harvested. Small output levels of respondents were attributed mainly to scarcity of land.

Table 12. Farm size and output level

Farm size (acres)	Output level		Total
	1 - 10 bags	11 bags and above	
1-5	57	12	69
6 and above	137	29	166
Total	194	41	235

Landlessness, featured prominently as a problem hindering the smooth operation of the daily economic activities of the inhabitants. To a question as to what were the major problems hindering their economic activities, 40 percent of the respondents complained of scarcity of land in their farming activities while 50 percent of the respondents attributed poverty to landlessness.

Land as the major productive asset in the study area was limited in supply. Land ownership was also limited to a few individuals who were lucky to be part of the early beginnings of the settlements. Apart from the few individuals who own land, the rest of lands are lineage and stool property. This gives greater access of land to the family heads, traditional authorities and the "Odikros". This arrangement in land ownership left most people with limited access to land. Because land was a lineage property, family members had only usufructuary rights but not ownership rights. This, coupled with the pressure put on the land through the ever-increasing population, has resulted in small farm sizes. The main form of land acquisition, which is through inheritance, suggests that land is subdivided after inheritance and children become poorer than their parents were because they are unable to engage in any meaningful production which would move them out of the poverty bracket.

The scarcity of land has resulted in exploitation of the would be farmers i.e., the youth by the land owners, family heads who hold lands in trust of family, and the "odikro" who is in charge of the stool lands. Lack of access to land has forced most of the youth into the traditional "abunu" and/or "abusa" system. The landlords release the land to the farmer but the preparation of the farm and its husbandry are left in the hands of the farmer. The farm produce is shared according to the tenants of the contract agreement. But the question is how many acreages could be put under cultivation by the farmer to sustain himself/ herself let alone making savings out of his/her share of the farm produce? The farmer, after a season's work, may not even cover the cost of production let alone make profit. The plight of the farmer is worsened when he/she is unfortunate to meet emergencies. Such emergencies include payment of hospital bills of immediate family members, payment of school fees and funeral expenses. In such instances, the farmer who engages in "abunu" or "abusa" is forced to consult his landlords for a loan or sell his portion of the farm which is yet to be shared at a give-away price. The farmer after meeting his/her emergency with his portion of the farm is left penniless most of the year. It is therefore, very difficult for the farmer who engages in shared-cropping to move out of poverty.

The scarcity of land may be due to increasing population because the land is fixed in supply but the population keeps on increasing day in and day out. This confirms [67] assertion that population growth has a lot of implications which may result in poverty, and [68] view that small farms are subdivided on inheritance and children become poorer than their parents were. This subdivision of land is due to high rate of population growth and its attendant pressure on land. According to [69],

> Uncontrolled population growth and uncontrolled exploitation of natural resources combine in a vicious circle. Thus, the more people there are, the more they destroy the long-term potential of fragile environments, and the poorer this makes them and their descendants.

People get more income as they put large tracts of land under cultivation. One hundred respondents constituting 59.6 percent of those who earned between one hundred Ghana cedis and approximately five hundred Ghana cedis had to cultivate 6 acres or more whereas only 41 constituting 57.7 percent who were in this income bracket cultivated between 1 – 5 acres (Table 13).

Table 13. Income and number of acres farmed

Income (GH¢)	1 – 5 acres		6 acres & above		Total	
	Freq	%	Freq	%	Freq	%
100 – 299.9	26	36.6	69	40.8	95	39.6
300 – 499.9	15	21.1	33	19.5	48	20.0
500 and above	30	42.3	67	39.6	97	40.4
Total	71	100.0	169	100.0	240	100.0

With those who earned 500 Ghana cedis and above 67 (39.6%) out of the 97 who were in this bracket cultivated more than 6 acres before they were able to accrue such an amount. Furthermore, respondents who earned between one hundred Ghana cedis and two hundred Ghana cedis cultivated more land as 40 percent cultivated six acres and more. This means one's income level depends on the size of land he/she puts under cultivation. However, this relationship was further subjected to proof and a chi-square was calculated to determine whether respondents' income level was related to the land sizes they cultivate.

From Table 14 a calculated χ^2 of 0.372 showed there was no relationship between income levels and sizes of farms. Therefore the increase in respondents cultivated acres of land did not depend on the incomes of respondents. The relationship between income level and farm size is further illustrated in Table 14.

Table 14. Farm size by income level

	Income (GH¢100)					
Farm size	Less than 1	1 to 2	2 to 2.99	3 to 3.99	4+	total
1 to 5	10.2	5.5	1.6	2.4	5.5	25.2
6 and above	29.1	29.1	2.4	2.4	11.8	74.8
Total	39.4	34.6	3.9	4.7	17.3	100

N = 250.

GENDER AND POVERTY

In most rural communities in developing countries, gender is associated with poverty. According to [70], women are more likely to be poorer than men. One way to establish the relationship between gender and poverty was through education and the following question was raised: how does access to education and gender determine the level of poverty in the study area? The data collected confirms that there was a relationship between sex and educational status. Whereas 36.6 percent of those who had no level of education were males, that of females was 63.4 percent. The male/female ratio changes in favour of males as the level of education increases. From the data, 50.3 percent of the male respondents had primary/Junior Secondary Education as against 49.7 percent of females.

The number of females decreases further as one moves up the educational ladder. In terms of secondary education, 85 percent of males had access whilst only 15 percent of females had access. The inability of females to have equal access to education as males could be buttressed by the following narration by an old lady at Hweehwee.

> The traditional Ghanaian culture which does not place value on female education and the notion that the place of women is the kitchen are the causes of poverty. My siblings (brothers) were schooled at my expense. Whilst they were schooling I was with my mother on the farm and engaging in other household chores. Parents, especially fathers in most traditional societies prefer sending the boy-child to school to sending the girl-child to school. Females are also seen as those who would ensure the continuous survival of the family by way of procreation. I was forced to marry at the age of sixteen simply because my mother wanted to see her grandchildren before she dies. This attitude has prevented most females in our traditional area from having access to formal education. I was not educated to attract any meaningful well-paid job, so I have been farming since infancy. It is therefore not surprising that my brothers are better off economically. The lack of formal education place most women in lowly placed jobs which attract lower wages. Also lack of formal education results in most women not being employable in the formal sector. They engage in farming and other self-employed occupations which do not guarantee a regular source of income. It is therefore not surprising that my brothers are better off economically. I am one of such women who are deprived access to education hence most women due to lack of education have lower and irregular incomes and permanently find themselves in the poverty trap.

The above narration confirms the idea that one's sex determines his/her access to education which tends to determine one's income. This also supports [71] view that women are likely to be poorer than men. As respondent remarked:

A major role for women is to ensure the continuity of the lineage, and she was expected to marry soon after puberty. Women do not need formal education to perform this function. A women was expected to be provided for by her husband, and since education became a means for entering highly-paid jobs in the formal sector, it was considered more important for boys to have formal education, since they were to be breadwinners in the family

Another respondent shared her experiences

Women are exploited a lot by men in this community which keeps most women in poverty. The exploitation is in the form of division of labour based on sex. My husband after clearing a piece of land for farming leaves the cultivation and other farm husbandry for me. I am expected to take good care of the farm and use the proceeds for housekeeping especially in terms of feeding the family. He would be generous to help in the farm husbandry but would not provide money for housekeeping since I am expected to do that though the farm proceeds. Women who would be lucky to get husbands to provide them with "seed" capital with which they would start petty trading are expected to feed the family out of the profits on their trading activities. Whereas I was taking charge of small household expenditures, my husband was saving money which was used to build a house in his name, (such property could be easily used as a security to secure loans from financial institutions). It is therefore not surprising that rural women are unable to make enough savings which would help propel them out of the poverty bracket.

The reasons why most women find themselves in poverty could also be traced to their exploitation by their fellow women i.e., the so called "market queens" in connivance with a few local elites who act as agents of the "market queens". They buy cheaply the products/produce from the rural women and sell to the urban dwellers at higher prices (The profit that would have gone to the rural women goes to the "market queens" and their agents). The "market queens" regulate the number of buyers who would go to the producing centres at a time to transact business with the people. They even at times refuse to go to the producing centres leaving their agents to do the purchasing. In such instance, they dictate the prices that would be paid to the producer.

Access to Inputs and Rural Poverty

The effects of the quality and quantity of inputs used and the level of output would also determine the level of productivity. That is what is invested or used in production determines what is harvested. To ascertain the relationship between input and output the following question was posed. Does the amount of inputs (i.e., fertilizers, insecticides, credit, labour, etc.) a farmer uses influences his/her level of output?

To find out the relationship between input and output levels, questions relating to the use of inputs were asked. The results showed that only 20.4 percent of the respondents had applied fertilizer in one time or the other, while 79.6 percent of the respondents had never used fertilizer before. It was also revealed from the studies that only 8.6 percent of the respondents had used insecticides before. The farmers did not use fertilizer and/or insecticides only because they did not like them, but they complained of high cost of these inputs.

Table 15. Output and use of fertilizer

Use of Fertilizer	Output (bags)	
	1-10	11 and above
Yes	51	16
No	142	24
Total	193	40

The farmers attributed low output levels to loss of soil fertility and the inability to apply fertilizer to improve the quality of land. The low level of output was further revealed in a cross tabulation as shown in Table 15. The inability to have access to inputs to facilitate large scale production to get enough income corroborates the view of [40] (p.37). In his view:

> The small farmer is unlikely to possess sufficient capital or financial resources and is hence unlikely to have access to improved farming methods which require fertilizers, chemical sprays, machinery etc. the major resources of the small farmer are land and the labour which can be provided by the household. Lack of asset holding prevents small scale farmers from using modern inputs of farming and inability to contract credit to help them finance the purchase of such inputs.

ACCESS TO CREDIT AND OUTPUT LEVEL AND RURAL POVERTY

Generally, lack of credit opportunities for cultivation is another cause of rural poverty. Mostly, rural areas lack credit facilities offered by banks or credit unions or micro credit offered by government and/or private sector organisations. In instances where rural farmers happen to obtain some form of credit facilities and are able to expand their farms, they often do not have available markets for their farm produce, thus leading to post harvest losses. Since most rural dwellers work largely in the agriculture sector, this incidence of non-existence of credit facilities coupled with post-harvest losses due to unavailability of markets for farm produce is a serious setback to their efforts at moving out of the poverty trap. The data collected revealed that the use of credit was not part of the culture of the people in their daily economic activities. It was detected that only 5 percent of the respondents had once used, or ever accessed credit in their economic activities.

The lack of use of credit by the farmers was not due to the fact that they did not want the credit but access to it was the problem. Lack of financial institutions in the study area might have accounted for the inability of the people to get access to credit for their economic activities. The nearest financial institution in the study area is about 30 kilometres away. Nkawkaw, which is the nearest urban settlement has ten banks with five of them being rural banks with none sited in the study area. The rural banks seem to have deviated from their primary function of assisting rural dwellers with low interest loans. Bankers prefer assisting urban based traders who have quick turnovers, and salaried workers who can use their monthly salaries as collateral securities. This deviation caught the attention of the then Governor of the Bank of Ghana who directed that all agencies of rural banks in the cities should be closed down by 31st December, 1998 [72].

These issues of lack of credit and inputs were identified as some of the main obstacles in the economic activities of the people in the study area. Seventy percent of the respondents indicated that the major problem encountered in their economic activities was lack of credit, while 65 percent listed lack and high cost of inputs as another setback. These obstacles do not allow the people to engage in any meaningful or large scale production to generate income which could move them out of poverty. When the respondents were asked what in their opinion, was the main cause of poverty in the area, 65 percent attributed poverty in the community to lack of credit facilities or financial support.

Inadequate credit or financial support as one of the main factors responsible for poverty in the study area is in agreement with view of [73] which states that the main factors that cause poverty in Sub-Saharan Africa include: inadequate access to employment opportunities; inadequate physical assets such as land, capital and minimal access by the poor to credit even on a small scale.

Chapter 7

MANIFESTATIONS OF LAND DEGRADATION

Another important objective set out in this study was to examine the manifestations of land degradation and factors responsible for land degradation in the study area. This chapter therefore, examines the manifestations and causes of land degradation in the Kwahu South District.

Evidences of land degradation such as non-availability of non-traditional forest products, rainfall patterns, cutting of trees for timber, vegetative cover, soil fertility, yield per acre and farm size as well as income level were examined.

NON AVAILABILITY OF NON-TRADITIONAL FOREST PRODUCTS AS MANIFESTATION OF POVERTY IN RURAL AREAS OF GHANA

According to [74] and [75], it is necessary to maintain the biodiversity of the existing forest ecosystem not only as an end of itself, but as a major requirement in the conservation of Non Traditional Forest Products (NTFPs), for their importance in the social, medicinal and dietary needs of the rural and urban population.

Both on-and off – reserves are rich in biodiversity. Over 2100 plant species have been found in the High Forest Zone in Ghana. A total of 730 tree species have been recorded from the closed forests. The Wet Evergreen forest is floristically the richest. The Moist Evergreen and Moist Semi – deciduous forests are the most important for commercial timber species [76]. The fauna of the forest zone includes over 200 species of mammals many of which are rare, such as endangered bats, rodents, a variety of reptiles and birds [77].

Forest with its biodiversity contributes to all aspects of rural life. It provides food, fodder, fuel, medicine, building materials, household items and income. Also important are intangible benefits such as cultural symbols, ritual artefacts and sacred sites [78]. The commonly traded NTFPs in Ghana include foods such as snails, mushrooms, fruits, seeds, bee honey and bush meat. Others are spices, chewing sticks or sponges, kola nut, medicines, canes/rattan and other household goods like mortar, pestles, calabash, sponges, and food-wrapping leaves.

However, the situation in the study area is quite different. There is the absence or general decline in the availability of NTFPs. Hardly do people come across some of these NTFPs in the forest, unless one travels kilometres deep into the forest. The disappearance or scarcity of the NTFPs is a clear indication of land degradation. People have to spend hours in fetching for firewood and water for domestic use. Quality man hours that would have been used for productive activity to generate income are lost thereby pushing those who depend on the forest into poverty. To maintain the biodiversity of the forest ecosystem as a major source of Non Traditional Forest Product (NTFPs), there is the need to curb land degradation to secure the abundance of these non-traditional forestry products.

5. The forested nature of the study area is not all lost as evidence of some forest reserve can still be seen as depicted in plate 6.

The plantain farm in Plate 7 is a clear indication of a degraded environment. The soil is not fertile enough to support plant life hence, the smaller sizes of the plantain suckers and the withered leaves. It would take a longer time for the plantain suckers to bear fruits due to the poor quality of the soil as compared to the plantain farm on Plate 8. Though there is no scientific soil test, indigenous knowledge maintains that the long gestation period for crops such as those in Plate 8 is an ample demonstration of loss of soil fertility.

Plate 5. A deforested and degraded land at Esaase.

Plate 6. A forest at Esaase.

Plate 7. A plantain farm on an unfertile land at Esaase.

Plate 8. A plantain farm on a fertile land at Esaase.

RAINFALL PATTERN AS MANIFESTATION OF POVERTY IN RURAL GHANA

The rainfall pattern in the study area between 2000 and 2004 further shows the manifestation and the extent of land degradation in the area. Table 16 compares the rainfall pattern from 2000 and 2004 in the Kwahu South District.

Table 16. Monthly total rainfall (mm) from 2000 to 2004

Months	2000	2001	2002	2003	2004
Jan	47.5	0.0	13.2	10.7	80.5
Feb	0	0.0	18.2	30.3	76.8
Mar	112.5	218.9	156.9	44.2	111.6
Apr	122.9	131.9	123.8	188.4	107.8
May	119.8	105.6	203.6	138.4	115.1
Jun	353.5	170.7	263.5	145.1	103.4
Jul	163.9	21.4	136.5	48.9	100.8
Aug	61.8	11.1	101.4	42.8	138.3
Sep	191.4	159.0	140.6	136.3	271.7
Oct	210.9	54.9	199.2	104.1	162.6
Nov	10.9	72.8	29.5	92.7	49.4
Dec	0.8	18.7	25	23.8	57.7

Source: Ghana Meteorological Agency (2002).

The figures in Table 16 showed a consistent decline in the average amount of rainfall recorded over the period. With the exception of the month of June 2002, the average rainfall for the peak months of June and July were declining. For instance the average rainfall for the month of June 2000 was 353.5 mm whereas that of 2004 was 103.4mm. This shows the extent of deforestation in the area since inadequate rainfall is a characteristic of deforestation and for that matter land degradation.

Chapter 8

Causes of Land Degradation

There were a number of factors responsible for land degradation in the study area. Among them were bush fires, cutting of trees for timber and cutting of trees for charcoal.

Bush Fires as a Major Cause of Soil Degradation

Bush fires have been identified as one of the major causes of land degradation hence reduction in output levels of farmers in the study area. Bush fires have destroyed the forest cover of the study area. The once evergreen forest has now turned into grassland and savannah.

Though it was very difficult to pinpoint the exact causes of the perennial bush fires, they were attributed to long periods of drought, resorting to bush burning to clear the land for farming purposes, setting fire to hunt for game, the activities of palm wine tappers, and intentional destruction of forest reserves for the forest/land to be allocated to them for farming purposes. Farmers may as a result of this lose all that they have invested that farming season. Those who might have contracted loans from banks and friends are not able to pay off their debts. Year in and year out, people lose their cocoa and other cash crop farms as a result of bush fires. Such cash crops are life time investments and it becomes very difficult for farmers who go through such experiences to get back on track.

Bush fires have destroyed fauna and flora which were used to supplement rural incomes. Even though people might have low incomes, access to Non Traditional Forest Products such as snails, mushrooms, fruits, edible leaves among others were more than enough for them to have balance diets. Due to bush fires resulting in the destruction of the forest, rural dwellers could no more get access to fruits and other forest products which were very common in the forest. These NTFPs were accessed freely by the rural dwellers. Therefore, the money that could have been used to buy them was saved or used for other purposes which could improve their poverty situation.

Bush fires may have direct effect on output levels by destroying the crops even before they are harvested. Farmers as a result are unable to pay off loans contracted for farming. Year in and year out, people lose their cocoa and other cash crop farms as a result of bush fires. Such cash crops are life time investments and it becomes very difficult for farmers who go through such experiences to get back on track. Sustenance becomes a problem and they are further plunged into poverty. Similarly, bush fires can result in the destruction of the vegetative cover of the land, thereby exposing the land to the vagaries of the weather. The soil for farming is therefore exposed to rainfall resulting in soil erosion which eventually washes away the top soil. The soil is also affected by the direct rays of the sun. This results in hardening of the top soil thereby affecting the soil fertility. Either way, output levels of the farmer would be reduced.

For instance, in an answer to a question on whether the vegetation in the area has been destroyed as a result of bush burning, 172 (69.4%) of the respondents were of the view that indeed the area's vegetation has been affected seriously by bush burning. This supports the assertion that bush burning is a serious cause of land degradation as it often leads to the destruction of relevant vegetation cover. Bush fires as a cause of land degradation in the study area is given prominence by [79] and [80]. The greatest direct threat to the long-term survival of Ghana's forest is wild fire which has destroyed about 30% of drier forests especially in the transitional zone. The extent to which bush fires have destroyed the forest cover of the area is shown in Plates 9 and 10.

Plate 9. Burnt forest at Onyimso.

Plate 10. A burnt plantain farm at Onyimso.

CUTTING OF TREES FOR TIMBER AS A CAUSE FOREST DEGRADATION

Another cause of poverty identified was cutting of trees for timber and energy. This was mainly done by chain saw operators. Apart from the problem of deforestation with its attendant effect of inadequate rainfall, there was also destruction of farms by the felled trees. The chain saw operators normally work at night therefore, even if they destroy farms, the farmers are unable to locate them for compensation. The destruction of the forest through cutting of trees has resulted in inadequate and erratic rainfall pattern in the area. About 75 percent of respondents claimed that their farming activities were being affected by poor rainfall. The rains do not come at the right time hence, the farmers are unable to predict and plant their crops to meet the raining season. Most farm crops especially cereals wither as a result of prolonged dry seasons and the erratic rainfall patterns. Farmers are therefore unable to recoup what they invest, let alone make profit hence, they remain poor.

The means of 1.61 and 1.67 for bush fires and logging respectively relate to a question as to whether respondents agreed or disagreed to certain factors as causes of land degradation. These means suggest that the two factors are part of the causes of forest degradation in the area. The assertion by [78] and [79] that a major cause of deforestation in Ghana is illegal or indiscriminate logging to meet the ever-increasing local and external demand for wood products collaborates this finding. About 48 percent of the 250 respondents who answered questions on factors that cause land degradation in the area revealed that logging and cutting of trees for timber was a major contributor to the land degradation in the area. Another 38 percent agreed that land degradation was due to cutting down of trees for energy use. The incidence of logging does not pertain to the study area alone. In 1992 alone, Ghana produced 1,330,000m^3 of logs, 1,148,000m^3 were sold locally and 182,000m^3 exported [78].

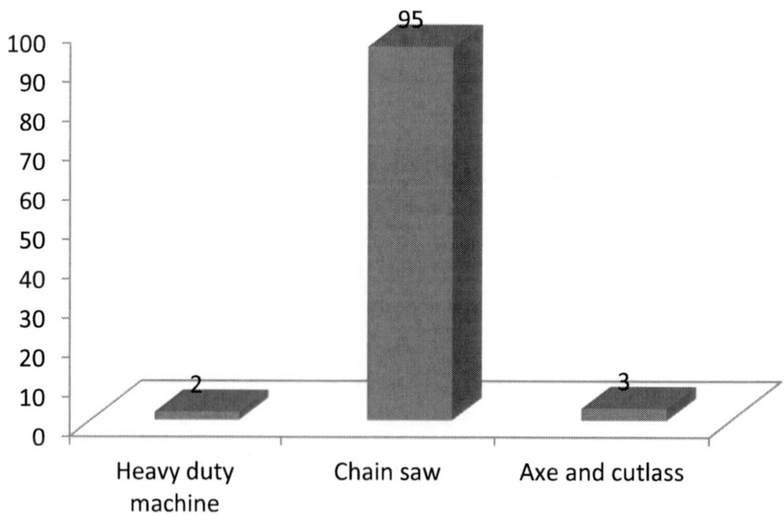

Figure 5. Mode of harvesting trees from the forest.

Figure 5 shows that the most widely used means of harvesting timber in the study area was chain saw. Ninety-five percent of respondents asserted that chain saw is used whereas only 2 percent talked of the use of heavy duty machines.

[81], reports that about 80 percent of Ghana's forests have been destroyed by illegal logging. This has resulted in more trees being harvested to meet the demands of local consumers. As more trees are harvested, the more the land is degraded through deforestation with its consequential effect of exposing the top soil to the vagaries of the weather, that is, rainfall and the scorching sun.

Not only was the forest depleted by the felled trees, the timber trucks and tractors used in carting the logs from the operational sites to the feeder/main roads most often destroyed the vegetative cover by way of constructing accessible roads resulting in deforestation. The land therefore becomes exposed to the vagaries of the weather.

Even though lumbering has been identified as a good economic opportunity as it provides some form of foreign exchange earnings and income for the country and the people respectively, the problem however was that, in most instances the contractors who often engaged in this enterprise did not have any serious concerns for the replacement of the trees felled. Even in cases where some form of replanting was done, no serious mechanisms were put in place to protect the young trees. This claim above was supported by data provided by the respondents on the question that sought to elicit their views on whether trees felled were often replaced by the contractors. Ten percent of them agreed that some form of forestation was done by the contractors whenever they felled trees in the area. However, 73.2 percent of them indicated that the contractors who use chain saws, axe and heavy duty machines in felling of the trees, did not often replace the felled ones through replanting of young seedlings. The effects of this act were serious deforestation and a consequential influence on rainfall patterns and farming in general.

Cutting Trees for Charcoal

Cutting trees for charcoal burning was also identified as one of the causes of land degradation in the area. Charcoal burning is a major economic activity of the youth in the study area. The high cost of Liquefied Petroleum Gas (LPG) and the inability of the people to afford it have compelled the people to use firewood and charcoal as sources of energy. In situations where they could afford LPG accessibility became a problem. The nearest LPG service centre to the study area is Nkawkaw which is about 30 kilometres away. The people therefore have no option than to resort to the use of charcoal and firewood as source of energy for cooking. It was therefore not surprising that 95.4 percent of households used charcoal and firewood as sources of energy for cooking. Plate 11 shows track loads of charcoal been transported to neighbouring town for sale.

Plate 11. Cargo trucks carting charcoal to the urban centres.

Plate 12. Charcoal being transported from Hweehwee to Nkawkaw.

This is a daily spectacle in the study area. A lot of cargo trucks are seen daily with loads of charcoal being transported to the urban centres for the energy needs of the people. This is further evident in plate 12.

This has very serious environmental consequence in the form of deforestation. Hweehwee and its environs supply the bulk of charcoal consumed in most of the urban centres in the Kwahu South District. The use of the forest as a source of energy in the study area is collaborated by the [82] which estimated that fuel wood consumption in 1988 was 15.9 million cubic metres and that this would rise to 17 million cubic metres by the year 2000. Table 17 sums up the factors responsible for land degradation in the study area.

Table 17. Factors that account for land degradation in the study area

Practice	SA	A	NA/D	D	SD	Mean	Std
Cultivation of a land continuously for more than five years	98	109	12	22	9	1.94	1.056
Burning the bush/vegetation cover as a means of land clearing	62	87	12	55	34	2.64	1.409
Using of tractor to clear the land	27	127	19	59	18	2.65	1.162
Slash and burn method of bush clearing	31	103	10	95	11	2.80	1.197
Shifting cultivation	33	87	18	81	31	2.96	1.304
Cutting of trees for charcoal burning	104	107	12	16	11	1.89	1.052
Cutting of trees for timber	120	91	9	20	10	1.83	1.083
Bush fires	135	94	4	7	10	1.65	0.954
Soil erosion	67	136	6	23	18	2.15	1.135
Pressure on land due to increase in population growth rate	60	115	8	44	23	2.42	1.278
Lack of adequate rainfall	83	128	5	20	14	2.01	1.086
Increasing mining activities in forest areas	140	71	5	7	27	1.84	1.282
Frequent dying of water bodies	93	108 9	14	20	2.08	1.251	

Chapter 9

RELATIONSHIPS BETWEEN RURAL POVERTY AND LAND DEGRADATION

INTRODUCTION

This chapter explains how the conceptual framework unearths the relationship between land degradation and rural poverty. The study has tried to establish a link between the key variables of poverty and land degradation and how these factors interact with each other in the study area.

The following conditioning factors were isolated for discussion: market prices, production and consumption technologies, information availability, population pressure, access to land and scarcity of land.

MARKET PRICES

Prices of goods and commodities are not set by the households but are determined by the external market prices or forces. Therefore, the household who farms on marginal lands and also produce low yields must buy essential commodities to survive. They also have the option to sell their produce at the price determined by the external forces hence forcing the household to engage in land intensification which could result in land degradation in the long run. Farmers may even be forced to sell at reduced prices to meet their needs and selling at reduced price means lower income hence poverty.

Mostly the prices of non- farm produce are set by institutional forces such as cost of raw materials, labour and other production costs whereas prices of farm produce are set by the interplay between demand and supply. Most produce are seasonal in nature. This results in bumper harvest at certain times. However, demand for food for instance is inelastic hence, supply most often outstrips demand. The need to meet basic needs of life coupled with inadequate storage facilities and lack of pricing policy for farm produce push farmers to sell all that they produce at reduced prices. Consequently, most rural farmers are not able to break even thereby perpetuating their poverty situation.

Technologies Used to Produce and Consume Natural Resources

The study area is largely rural in nature. In the case of this research the use of obsolete tools such as cutlasses and hoe means that there is land intensification as one piece of land is cultivated continuously over a long period. This would lead to loss of soil fertility hence low productivity. This further implies that with more marginal land under cultivation the tendency for degradation is higher. With the use of traditional farming methods such as slash and burn and mono cropping the land is subjected to intense pressure hence degradation at a fast rate. Farmers do not apply modern methods of farming, for instance, the application of fertiliser was not encouraging. In a question as to whether respondents applied fertiliser on their farms during the previous farming season, only 20.4% affirmed its usage. Some attributed the non-usage of fertiliser to the high cost of the inputs whereas others alluded to poor quality of farm produce.

Inadequate Access to Information

There is also the problem of inadequate access to information for farmers in the rural community. The unavailability of information may be due to the gap between research and practice. Research findings about appropriate agricultural technologies and modern methods of farming which may eventually improve rural agricultural productivity remain on the shelves of researchers and academics in the various institutions. There is also the issue of inadequacy of agricultural extension staff to disseminate valuable information to rural farmers. Therefore, these rural farmers resort to the use of only traditional and indigenous methods of production which may not increase productivity.

Population Pressure

Firstly, the study revealed that a large family size was a cause of poverty. Large family size was a common feature in the study area as the average family size of five members per household was recorded in the study area. The field data indicated that 60.8 percent of the respondents had more than five members in a household. This has consequences for land degradation since large families mean large tracks of land would be put under cultivation to increase yield to feed the family. High population growth rate and large family sizes are common features of most rural communities in Ghana with the study area being no exception. The large family sizes have resulted in most families in the rural areas living impoverished lives. This emanates from the fact that consumption levels of such families are higher and per capita income being lower. The large families depend on the land for survival. Considerable pressure is put on farm lands resulting in over cultivation, over intensification and extensification of land use. Most farm lands become barren as a result. Rural farmers are therefore forced on to marginal farm lands. This reduces the productivity of farm lands. The continuous cultivation on marginal lands leads to land degradation. Low productivity due to land degradation results in low income hence poverty.

Large family sizes emanate from adherence to old cultural practices, customs and traditional beliefs. For instance, some people justified the need for having more children on the ground that some of them would die. This may seem to be irrational at this age of medical science and improvement in the heath delivery system but their fears are somehow justified on the ground of lack of health facilities, lack of potable water and poor sanitation exposing both mothers and children to health hazards. Large family size was also justified by the people on the firm belief that one of them would be successful or rich in future. Children are also seen as an insurance against old-age. For parents to get someone to fall on in times of need (especially during old age and sickness) they decide to have more children so that cach would contribute to their upkeep when the need arises. Some men in the desire to collect the traditional "badu guan" would stop at nothing unless they have collected the ultimate prize. Men in particular see the number of children they possess as a source of pride and have been boasting about the number of their children.

Some rural dwellers also see children as gifts from the gods so they should not offend their ancestors and gods by using contraceptives or resorting to abortion.

The aforementioned factors which naturally would contribute to large family sizes perpetuate rural poverty. A greater part or almost all of what is produced by the family is consumed without selling some to generate income to meet other needs, especially health and educational requirements. Because of the large number of children, parents are unable to give their wards any appreciable level of formal education which may lead to securing alternative form of employment other than farming. It is therefore not surprising that, both income poverty and human poverty are very prevalent at the study area. Large family sizes most often result in farmers keeping or storing almost all of their farm produce for home consumption. Small farm sizes do not permit large scale production. Production is therefore geared toward home consumption. Producing for sale is rarely the case. What is produced is even unable to feed members of the family let alone selling some to get income to meet other needs of the family. Most farmers even purchase food to supplement what they are able to produce.

ACCESS TO AND SCARCITY OF LAND

Access to land/farm size was revealed as a cause of rural poverty and land degradation. Land as the major productive asset in the study area is limited in supply. Its ownership as a result of its limited supply is also limited to a few individuals who were lucky to be family heads. Apart from the few individuals who own their own land, the rest of the land is either lineage or stool property. This gives greater access of the land to the family heads, traditional authorities and the "odikros". This arrangement in land ownership leaves most people who are in the majority limited access to land. Due to the fact that land is a lineage property, family members have only usufructary rights but not ownership rights. The difficulties associated with land acquisition, the land tenure system and the small size of most rural farms coupled with the use of rudimentary agricultural techniques and practices can equally result in land degradation. Thirty-seven percent of the respondents reported that they had encountered some problems or difficulties in the acquisition of their farmlands. The nature of land acquisition and access cause households or individuals to take decisions that can result in land degradation. This is because access to land may result in cropping on one piece of land for a long time resulting in loss of soil fertility and easy run off of rain water with the slightest rainfall. This also leaves the surface of the land bare and with rainfall the eventual result is land degradation. Again this is a case of a relationship between poverty and land degradation that has been manifested in the conceptual framework of the study. The research has further demonstrated that in the study area there is land degradation as evident in the reduction of crop yield, loss in soil fertility, loss of vegetative cover and even rainfall pattern.

Furthermore, scarcity of land resulting in small farm sizes equally explains the relationship between land degradation and poverty. Lack of access to the land has forced most of the youth into the traditional "abunu" and/or "abusa" system. The landlord releases the land without contributing anything again in terms of investment either in labour, inputs or cash. Land preparation and farm husbandry is left in the hands of the poor farmers only for the farm to be shared when the crops bear fruits or the proceeds to be shared according to the tenants of the contract. But the question is how large is the size of the land that would be cultivated for the farm or proceeds to be shared. The poor farmer after a farming season's work may not even cover cost of production let alone making profit. The exploitative nature of the "abunu" and/or "abusa" contract pushes the poor farmer further into poverty while the rich gains more. This issue of scarcity of land leading to small farm sizes has been found to be one of the factors responsible for poverty in the study area.

LEVEL OF EDUCATION

Another important factor in explaining the relationship between poverty and land degradation is the level of education. Level of education reduces poverty as individuals climb up the educational ladder. On the other hand, low level of education or lack of education can equally contribute significantly to environmental and for that matter land degradation. This is because illiteracy means lack of knowledge of basic farming practices and activities that can help preserve the land for proper and maximum use. The situation in the study area as far as education is concerned deserves special attention as 92.4 percent of those who engage in agriculture had basic education as the highest level of educational attainment. Only 11.6 percent of the respondents had accessed secondary education and above. This can directly impact on the use of land in the area.

The World Bank attributes poverty in Sub-Saharan Africa to low endowment of human capital resulting from inadequate access to education. Findings of the study points out that low education manifests over reliance on agriculture, as there is little employment diversification. Given the limited knowledge that rural dwellers have about land management and best farming practices, they engage in farming practices that easily lead to land degradation such as bush burning to clear the land, slash and burn and over cultivation. All these render the land bare and hence susceptible to degradation.

The result from the field has revealed evidence of poverty and evidence of land degradation which are the result of an interaction of the conditioning factors and the decisions of household members. The evidence of poverty from the study area was manifested in the form of inadequate food intake, high incidence of adult illiteracy, lack of social amenities, and use of charcoal as source of firewood, low income level, poor quality of life and lack of household assets.

The study equally revealed that the environment has been degraded over the years with this degradation being manifested in the form of unavailability of non-traditional forest products, change in rainfall patterns, degradation of vegetative cover, loss in soil fertility and loss in yields per acre over a period of a decade. Other environmental consequences that the study revealed resulting from the use of the land and evident in the study area were bush fires, loss of soil fertility, insufficient rainfall, cutting of trees for charcoal and logging. Therefore, land degradation is the end result of these interactions of conditioning factors, causes of poverty, household decision and environmental factor.

Chapter 10

MAJOR CONCLUSION

The research concludes that low levels of education in the study area, large family sizes and land degradation were dominant factors that served as causes and manifestations of poverty.

It can also be concluded that bush fires, logging, charcoal burning and over cultivation as well as scarcity of non-traditional forest products, poor rainfall patterns, loss of soil fertility and loss of vegetative cover were some of the causes and manifestations of land degradation.

It is evident from the study that there was a relationship between land degradation and rural poverty and degradation results in low output levels which subsequently leads to reduction in income levels hence increase levels of poverty. On the other hand poverty results in the inability of land users to invest in their lands by way of adapting modern farming practises to improve upon the quality of the land and securing new and fertile lands for farming hence the continuous cultivation of a piece of land over a long period which results in land degradation. The relationship between the two variables is not one dimensional but rather each variable impact on the other. This means that land degradation leads to poverty and rural poverty also leads to land degradation.

It would be impossible to address the problems of land degradation and poverty separately to achieve any effective results. To achieve maximum results there is the need to address them jointly.

References

[1] Kanwer, J.S. 1982. *Managing Soil Resources to meet the Challenges of Mankind.* Presidential Address, 12th International Congress of Soil Science, New Delhi.

[2] UN-REDD. 2011. The United Nations Collaborative Program on Reducing Emissions from Deforestation and Forest Degradation in Developing Countries. (Website: http://www.un-redd.org/)

[3] Food and Agriculture Organisation 2000. *Land resource potential and constraints at regional and country levels.* World Soil Resources Reports 90.

[4] World Resources Institute 1990. *Tropical Forests.* The World Bank and the United Nations Development Programme, Washington, D.C.: WRI.

[5] UNEP 2001. An Assessment of the Status of the World's Remaining Closed Forests. UNEP/ DEWA/TR 01-2. Division of Early Warning and Assessment (DEWA) United Nations Environment Programme (UNEP) P.O. Box 30552, Nairobi, Kenya.

[6] Reid, W. V., & Miller, K. R. 1989. *Keeping Options Alive: The Scientific Basis for Conserving Biodiversity.* Washington, DC: World Resources Institute.

[7] Food and Agriculture Organisation 2001. *Poverty, Food Security, and Agricultural Biotechnology: Challenges and Opportunities.* Report of the food and agriculture organization of the United Nations, FAO.

[8] United Nations Educational, Scientific, and Cultural Organization (UNESCO). 2007. Education for all global monitoring report: Reaching the marginalized. Paris: UNESCO.

[9] Serageldin, I. 1992. Governance and the Development Process. *Finance and Development, 28* (3), 13–24.

[10] Mayers, James. "Trees, Poverty and Targets:Forests and the Millennium Development Goals." IIED, 2007.

[11] Newman, L. 1998. *Qualitative-quantitative research methodology exploring the interactive continuum.* Southern Illinois: Carbondale, IL (pp. 61-74), Washington, DC: Island Press.

[12] Green, C. P. A. 1992. *The environment and population growth decade for Action.* Population Reports Series M, No. 10. Population Information Programme, Baltimore, USA: John Hopkins University.

[13] Tolba, M. K. 1993. *The Earth Summit and Africa's Development.* Ibadan, Nigeria: IITA.

[14] Food and Agriculture Organization 2007. *Land use systems of the World* www.fao.org. Accessed 15 June 2008.

[15] World Bank 2002. The Environment and the Millennium Development Goals. Washington, DC: World Bank. Online at http://www.wds.worldbank.org/servlet/

[16] Lele, U. and S.V. Stone. 1989. Population pressure, the environment and agricultural intensification, variations on the Boserup hypothesis. Managing Agricultural Development in Africa (MADIA) Discussion Paper. The World Bank, Washington DC, USA.

[17] Boserup, E. 1981. *Population and Technological Change.* Oxford, United Kingdom: Blackwell. 170-230.

[18] Ruthenberg, H. 1980. *Farming Systems in the Tropics.* 3rd ed., Oxford: Clarendon Press.

[19] Atongi, A. 1998. *Population, poverty and environmental degradation in the Bawku East District,* Unpublished Master of Arts dissertation. University of Cape Coast, Cape Coast.

[20] Morvaridi, B. 1994. Management of water supply and sanitation projects in Maharashtra State, India. *Journal of International Development, 6,* 129–132.

[21] Schnaiberg, A. 1980. *The environment: From surplus to scarcity.* New York: Oxford.

[22] Durning, A. B. 1990. *Actions at the grassroots: Fighting poverty and environmental decline.* World watch paper 88, World watch Institute, Washington D.C.

[23] Stonich, S. C. 1989. *Land degradation and society.* London: Methuen.

[24] Food and Agriculture Organization 2006. *Global forest resources assessment.* Forestry paper 147 FAO, Rome.

[25] Nsiah-Gyabaah, K. 1994. *Environmental degradation and desertification in Ghana.* England: Avebury Ashgate Publishing Ltd.

[26] Ministry of Lands and Natural Resources. 2011. Revised forest and wildlife policy of Ghana. Pp 1-38.

[27] World Bank. 2008. *High-level Event on the Millennium Development Goals.* New York: United Nations Headquarters.

[28] United Nations Development Programme 2001. *GLOBIO. Global Methodology for Mapping Human Impacts on the Biosphere.* UNEP/GRID-Arendal.

[29] UNDP 2000. *Human Development Report 1990.* New York: Oxford University Press.

[30] UN Millennium Project, 2005. Investing in development: A practical plan to achieve the Millennium Development Goals. Report to the UN Secretary General.

[31] International Fund for Agricultural Development, 2001. *Rural Poverty Report: The Challenge of Ending Rural Poverty.* Oxford University Press.

[32] Abdulhadi, R., Kartawinata, K. & Sukardjo, S. 1981. Effects of mechanised logging in the lowland dipterocarp forest at Lempake, East Kalimantan. Malaysian Forester, 44, 407–18.

[33] World Bank, 2001. *Understanding poverty.* www.worldbank.org.

[34] Ghana Statistical Service 2000. *Pattern of Rural Poverty in Ghana.* Accra, Ghana Statistical Service.

[35] UNDP (United Nations Development Programme. 2005. Comments on The Concept Note: Joint World Bank and IMF Report on Poverty Reduction Strategy Papers- Progress in Implementation, 2005 PRS Review. New York.

[36] Ghana Statistical Service. 2008. *2003 Core Welfare Indicators.* Accra: Ghana.

[37] IMF. 2005. "Strengthening the fund's ability to assist low-income countries meet balance of payments needs arising from sudden and exogenous shocks." Draft Washington, D.C.

[38] Malik, S. J. 1998. *Rural poverty and land degradation: What does the available literature suggest for priority setting for the Consultative Group on International Agricultural Research?* A report prepared for the Technical Advisory Committee of the Consultative Group on International Agricultural Research. Vienna, Virginia, Draft, February.

[39] Webster, Andrews 1990. *Introduction to sociology of development.* London: MacMillan Publishers.

[40] UNECA-SA (United Nations Economic Commission for Africa, Subregional Office for Southern Africa). 2005. Economic and Social

conditions in Southern Africa 2003: The Challenge of Private Sector Development in Southern Africa. Addis Ababa, Ethiopia.

[41] Townsend, P. 1979. *Poverty in the United Kingdom, a Survey of Household Resources and Standards of Living*. London: Penguin Books and Allen Lane.

[42] North, P. J. 1987. *People in society*. England: Longman Group Ltd.

[43] Sancho, A. (1996). *Policies and programmes for social and human development*. A handbook produced for United Nations World Summit for Social Development. International Center for Economic Growth, San Francisco.

[44] Olayemi, J. K. 1995. *A Survey of approaches to poverty alleviation*. A paper presented at the NCEMA Workshop on Integration Poverty Alleviation Strategies into Plans and Programmes in Nigeria, Ibadan.

[45] Institute of Statistical, Social & Economic Research (ISSER). 1970. Policies and options for Ghana's economic development. Institute of Statistical, Social & Economic Research (ISSER) University of Ghana, Legon, Ghana.

[46] Vizard, P. (2001). Poverty and human rights: Sen's 'capability perspective' explored. Oxford: Oxford University Press.

[47] Nussbaum, M. C. 2003. Capabilities as Fundamental Entitlements: Sen and Social Justice. *Feminist Economics*, *9* (2 & 3), 33-60.

[48] World Bank. 2007. World Bank report on Poverty. New York. Oxford University Press.

[49] Oldeman, L.R., R.T.A. Hakkeling & W.G. Sombroek. 1991. World Map of the Status of Human-Induced Soil Degradation: An Explanatory Note. ISRIC (International Soil Reference and Information Center), Wageningen, The Netherlands. (Gives a numerical overview of the importance of land degradation, classified per type of degradation and per specific phenomenon at world and individual continental level).

[50] The United Nations. 2000. The Millennium Development Goals Report 2000. New York: The United Nations.

[51] Eswaran H, Lal R, Reich PF. 2001. Land degradation: An overview. In: Bridges, EM, Hannam ID, Oldeman LR, Pening de Vries FWT, Scherr SJ, Sompatpanit S, eds. Responses to Land Degradation. Proc. 2nd. International Conference on Land Degradation and Desertification.

[52] Barrow, C. J. 1991. *Land degradation, development and breakdown of terrestrial environments*. Cambridge: Cambridge University Press.

[53] Wim, G., & El-Hadji, M. 2002. *Causes, general extent and physical consequence of land degradation in arid, semi arid and dry subhumid*

areas. FAO, Rome: Forest conservation and natural resources, forest Dept.

[54] United Nations Environment Programmes 1992. *World Atlas of Desertification*. London: E. Arnold.

[55] United Nations. 1978. Round-up Plan of Actions and Resolutions. United Nations Conference on Desertification, 29 August– 9 September, 1977, 43 pp. New York: United Nations. [This is the official report of the UNCOD.]

[56] Nkansah- Kyeremateng, K. 2000. *Kwahu Handbook: Tips galore for investors and tourist.* Bepong, Ghana: Sebewie Publishers.

[57] Agbesinyale, P. K. 1992. *Agricultural Expansion and the State of Ghana's Environment The need for Sustainable Development.* Discussion Paper Series No.2, Centre for Development Studies, U.C.C.

[58] Ghana Statistical Service. 2002. *The 2000 Population and Housing Census*. Accra: Ghana Statistical Service.

[59] Muhammed, M.A.S. 2007. Rural Development and Enterprise Development, Alexandria: Sustainable Development Association.

[60] Opoku-Mensah, Y. 2012. An analysis of human resource planning and its effect on organisational effectiveness – a case study of information services department, Accra office.

[61] Ghana Statistical Service (GSS). 2004. Ghana living standard survey 2004.

[62] Boserup, E. 1965. *The condition of agricultural growth.* London: Allen and Unwin.

[63] Ewusi K. 1978. *Plannning for the neglected rural poor in Ghana.* Legon: I.S.S.E.R. Technical Publication.

[64] Schiller, R. B. 1973. *The Economics of poverty and discrimination*. New Jersey: Prentice Hill Inc., Englewood.

[65] GPRS, 2004. *Ghana Poverty Reduction Strategy.* Accra.

[66] Glewwe, P., & Twum-Baah, A. 1991. *Distribution of welfare in Ghana 1987/1988.* Living Standard Measure study working paper no.75 Washington D.C.

[67] Ravallion M., & Sen, B. 1994. *Impacts On rural poverty of land based targeting, poverty and human resources division*. World Bank. Mimeo.

[68] Bloom, L., & Ottong, J. G. 1987. *Changing Africa: An introduction to sociology*. London: Macmillan.

[69] Bibangambah, R. J. 1985. Approaches to the problem of rural poverty in Africa. In Kiros, F. (Ed.). *Challenging Run fall Poverty*. Trenton, New Jessy: Africa World Press.

[70] Chambers, R. 1983. *Rural development: Putting the last first.* England: Longman scientific and Technical.

[71] Lipton, M. 1983c. Demography and Poverty, World Bank Staff Working Paper 623, Washington, DC: World Bank

[72] Dolphyne, F. A. 1991. *The emancipation of women: An African Perspective* Accra: Ghana, Universities Press. Khon Kaen, Thailand. Oxford Press, New Delhi, India.

[73] Obadan, M. 1997. Analytical framework for poverty reduction: Issues of Economic Growth Versus Other Strategies. Proceedings of the Nigerian Economic Society Annual Conference on Poverty Alleviation in Nigeria 1997. Ibadan: NES: 1-18.

[74] Thompson, I., Mackey, B., McNulty, S., & Mosseler, A. 2009. Forest resilience, biodiversity, and climate change: a Technical Series no. 43. Montreal, Canada, Secretariat of the Convention on Biological Diversity.

[75] Asibey, E. O. A., & Becko, C. Y. A. 1989. Forest & Food Security. In Owusu G. J. K. (Ed.). *Revision of Ghana's Forest Policy.* Accra: Forestry Commission symposia series no.3.

[76] Abbiw, D. (1990). *Useful plants of Ghana.* England: Intermediate Technology Publications Ltd.

[77] Hall, J.B., & Swaine, M.D. 1981. Geobotany –distribution and ecology of vascular plants in a tropical rainforest forest vegetation in Ghana. Dr. W. Junk Publishers. The Hague pp. 382.

[78] Kotey, E. N. A., Francois, J., Owusu, J. G. K., Yeboah, R., Amanor, K. S., & Antwi, L., 1998. *Falling into Place*, Policy that works for forests and people series No. 4. Study undertaken under the IIED, United Kingdom, and the MLF, Ghana.

[79] Ghana Forest Service. 1998. Manual of Procedures forest resource management planning in the HFZ. Unpublished. Forest Service. Accra.

[80] Swaine, M. D., V. K. Agyeman, B. Kyereh, T. K. Orgle, J. Thompson, & E. M. Veenendaal. 1997. Ecology of forest trees in Ghana. ODA Forestry Series. University of Aberdeen, UK.

[81] Marfo, E. 2010. Chainsaw Milling in Ghana: Context, drivers and impacts. Tropenbos International, Wageningen, the Netherlands. xii + 64 pp.

[82] World Bank. 1988. Staff appraisal report: Ghana Forest Resources Management Report. No. 7295-GH. World Bank, Washington, D.C, USA. 119 pp.

AUTHORS' CONTACT INFORMATION

Dr. Alex Obeng Somuah
University of Cape Coast
Senior Lecturer
Cape Coast, Ghana

Dr. Edward Debrah Wiafe
Presbyterian University, Senior Lecturer
P.O. Box 393, Akropong-Akuapem
Email : edward.wiafe@presbyuniversity.edu.gh

INDEX

A

B

C

D

E

F

G

T

U

V

W